CONTESTED FUTURES

W0113695

Contested Futures

A sociology of prospective techno-science

Edited by

NIK BROWN

BRIAN RAPPERT

ANDREW WEBSTER

Science and Technology Studies Unit (SATSU)
Department of Sociology, The University of York

Routledge
Taylor & Francis Group

LONDON AND NEW YORK

First published 2000 by Ashgate Publishing

2 Park Square, Milton Park, Abingdon, Oxfordshire OX14 4RN
52 Vanderbilt Avenue, New York, NY 10017

Routledge is an imprint of the Taylor & Francis Group, an informa business

First issued in paperback 2020

Copyright © Nik Brown, Brian Rappert and Andrew Webster 2000

All rights reserved. No part of this book may be reprinted or reproduced or utilised in any form or by any electronic, mechanical, or other means, now known or hereafter invented, including photocopying and recording, or in any information storage or retrieval system, without permission in writing from the publishers.

Notice:
Product or corporate names may be trademarks or registered trademarks, and are used only for identification and explanation without intent to infringe.

British Library Cataloguing in Publication Data
Contested futures : a sociology of prospective
   techno-science
   1. Social prediction 2. Twenty-first century - Forecasts
   I. Brown, Nik II. Rappert, Brian III. Webster, Andrew
   303.4'9

Library of Congress Catalog Card Number: 00-133531

ISBN 13: 978-0-7546-1263-6 (hbk)
ISBN 13: 978-0-367-60494-3 (pbk)

Contents

List of Contributors

Nik Brown is a Research Fellow in the Science and Technology Studies Unit (SATSU) at the University of York. His main areas of theoretical interest relate to the sociology of risk and identity, especially in the context of mammalian transgenics; the sociology of time and future-oriented coordination; Science and Technology Studies (STS) approaches to innovation dynamics in the Health and Life Sciences. He also serves as an advisor on a UK Government Health Task Force. Contact address: Dr Nik Brown, SATSU, Department of Sociology, University of York, York YO10 5DD UK. Tel. +44 (0)1904 434741. Email: ngfb1@york.ac.uk

Cecilia Cabello is a researcher at the Spanish Council for Scientific Research (CSIC). She holds a Masters degree in Science, Technology and Society, from the Universidad Autónoma de Madrid (Spain). She has worked for the European Commission JRC Institute for Prospective Technological Studies in Seville. Her areas of research include technology foresight and innovation; technology, employment and competitiveness; public policy in science and technology. Contact address: Consejo Superior de Investigaciones Científicas (CSIC), Unidad de Políticas Comparadas (former CSIC Institute for Advanced Social Studies) C/ Alfonso XII, n 18 (E-28014 Madrid, SPAIN), Tel. +34 915.219.160/028. Email: Ccabello@iesam.csic.es

J. Jasper Deuten obtained degrees in Technical Business Administration, and Philosophy of Science, Technology and Society at the University of Twente (Netherlands). He explored the social embedding of new technologies and product innovation management in the biotechnology industry. He is currently completing his PhD thesis on the emergence and stabilization of technological regimes. In addition to a sociological analysis of regimes, he is interested in the narrative form in technology development. Contact address: University of Twente, Centre for Studies of Science, Technology and Society, PO Box 217, 7500 AE Enschede, The Netherlands. Email: j.j.deuten@wmw.utwente.nl

Frank W. Geels studied Philosophy of Science, Technology and Society at the University of Twente (Netherlands) after which he conducted research for the Ministry of Transportation. His study analysed historical change in images of the future about the impact of information and communication technologies on transportation. He is now completing a PhD on long-term and large-scale technological transitions. He is also working on developing a new futures methodology for exploring technological transitions, namely socio-technical scenarios. Contact address: University of Twente, Centre for Studies of Science, Technology and Society, PO Box 217, 7500 AE Enschede, The Netherlands. Email: f.w.geels@wmw.utwente.nl.

Bastiaan de Laat is manager in the French arm of the science and technology consultancy, *Technopolis*. He works on programme evaluation and strategic management of R&D at institutional, national and international level. Recent projects include the mid-term evaluation of the French land transport research programme (Predit) and technology transfer to the Baltic Countries. Previously he worked at the Centre de Sociologie de l'Innovation of the l'École des Mines de Paris. Prior to that he was at the University of Amsterdam and also at TNOs Centre for Technology Policy Studies in the Netherlands. He holds an MSc in Chemistry from the University of Nijmegen and a PhD from the University of Amsterdam in 'Methodology for the strategic analysis of techno-economic networks'. Contact address: Technopolis France SARL, 5 rue de Castiglione, 75001 Paris, France. Tel. +33153 45 27 47. Email: bas.de.laat@technopolis.co.uk

Harro van Lente is Assistant Professor at the University of Utrecht (Netherlands). He graduated in physics and in Philosophy of Science and Technology at the University of Twente (Netherlands). In 1993 he received his PhD on the role of expectations in technological developments. After post-doc positions at the University of Oviedo (Spain) and Maastricht (Netherlands) he joined KPMG as a research manager of a strategic think tank. Currently, he is developing a graduate course in Science and Innovation Management at the University of Utrecht. Contact address: University of Utrecht, Centre for Science and Policy, PO Box 80068, 3508 TB Utrecht, The Netherlands. Email: h.vanlente@chem.uu.nl

Tom Ling is Reader in Public Policy at Anglia University, UK. He has published extensively around the issues of long-term public sector change and the transformations of the state. He has worked strategically with a wide range of public sector organisations during the past five years and is currently a member of the UK Government's Task Force on the future organisation of healthcare. Contact address: Sociology and Politics, APU, East Road, Cambridge CB1 1PT, UK. Email: tling@anglia.ac.uk

Mike Michael lectures in Sociology at Goldsmiths College, University of London. He has published extensively in a number of fields including the public understanding of science, critical social psychological theory, and the animal experimentation controversy. More recently his interests have turned to the role of mundane technologies in social ordering and disordering, and the role of social theory in the public understanding of science. He is author of *Constructing Identities* (Sage, 1996) and *Reconnecting Culture, Technology and Nature: From Society to Heterogeneity* (Routledge, in press). Contact address: Dr Mike Michael, Department of Sociology, Goldsmiths College, University of London, New Cross, London SE14 6NW, UK. Email: M.Michael@gold.ac.uk

Annemiek Nelis is a post doctoral research fellow in the Philosophy of Science and Technology, University of Twente (Netherlands). She obtained her Doctorate from the University of Twente in 1998 having addressed the socio-historical development of clinical genetics in the Netherlands. Since 1997 she has been working on the future development of genetic-diagnostics and IT in the health sector, first at the Science and Technology Studies Unit in Cambridge and, more recently, at the University of Twente. Contact address: University of Twente, Centre for Studies of Science, Technology and Society, PO Box 217, 7500 AE Enschede, The Netherlands. Email: a.p.nelis@wmw.utwente.nl

Brian Rappert is a Research Fellow in the Science and Technology Studies Unit at the University of York. His work has covered topics such as the commercialisation of university research, medical innovations, and research methodology. Currently, his research focuses on the social and political implications of 'non-lethal' weapons used by military, paramilitary, and police forces. Contact address: Science and Technology Studies Unit, Department of Sociology, University of York, York YO10 5DD, UK. Email: brr1@york.ac.uk

Arie Rip is Professor and Head of Department, Philosophy of Science and Technology, University of Twente (Netherlands). Educated in chemistry and philosophy at the University of Leiden (Netherlands), he shifted to science, technology and society studies. He has published extensively in the sociology of science and technology, in science policy studies and in management and assessment of technology in society. He is active in European Union research programmes and networks, and advises on science organization and science policy in a number of countries. Contact address: University of Twente, Centre for Studies of Science, Technology and Society, PO Box 217, 7500 AE Enschede, The Netherlands. Email: a.rip@wmw.utwente.nl

Hilary Rose is Visiting Research Professor in Sociology at City University, and Professor of Physic at Gresham College. She is also Emerita Professor of Social Policy at Bradford University. Her publications on science range from *Science & Society* (1969), co-written with Steven Rose, to *Love, Power and Knowledge: Towards a Feminist Transformation of the Sciences* (1994) and *Alas Poor Darwin: Arguments Against Evolutionary Psychology* (2000), edited with Steven Rose. She is currently preparing a book on the social dimensions of the new genetics. Contact address: Visiting Research Professor of Sociology, City University, 4 Lloyd Square, London WC1X 9BA. Tel. +44 (0) 207 713 1709.

Luis Sanz-Menéndez is Senior Research Fellow at the Spanish Council for Scientific Research (CSIC). Educated in Political Sciences and Sociology at the Complutense University of Madrid and at BRIE/IIS, University of California at Berkeley. He has researched and published extensively on science and technology policy and on the dynamics of research and innovation systems, from an institutionally based approach. He is involved in several European Research activities and represents Spain on different socio-economic research Committees of the EU. He is also on the Board of the EASST (European Association for the Study of Science and Technology). Contact address: Consejo Superior de Investigaciones Científicas (CSIC). Unidad de Políticas Comparadas (former CSIC Institute for Advanced Social Studies) C/ Alfonso XII, n 18 (E-28014 Madrid, Spain), Tel. +34 915.219.160/028. Email: Lsanz@iesam.csic.es

Wim A. Smit is Associate Professor of Science, Technology and Society (STS), at the University of Twente. After receiving his PhD in Physics (1973), he went into STS studies. He has published on risk assessment, assessment of nuclear technology, nuclear proliferation, and assessment and dynamics of military technological Developments. He has been a member of three Advisory Committees of the Netherlands Health Council: on External Safety; on Reassessing Nuclear Energy; on Risk Evaluation. His current research focuses on technological regimes, scripts and socio-technical networks in the field of Information and Communication Technologies. A second interest is in the increasing integration of civil and military technology. He has also been involved in the development of a methodology for designing socio-technical scenarios. Address: University of Twente, Centre for Studies of Science, Technology and Society, PO Box 217, 7500 AE Enschede, The Netherlands. Email: w.a.smit@wmw.utwente.nl

Andrew Webster is Professor in the Sociology of Science and Technology and Director of the Science and Technology Studies Unit at the University of York, UK. His main research interests relate to knowledge dynamics, the sociology of health innovations, and the commercialisation of public research. In 1999 he was appointed Director of the ESRC's Innovative Health Technologies Programme, a £5 million initiative exploring the social dimensions of health technologies. He serves on various national ESRC Committees, a Council member of European Association for the Study of Science, Technology and Society, and on the UK government's Post-Genomics Co-ordinating Committee. His recent publications include, *Capitalising Knowledge* (SUNY Press); *Innovation and the Intellectual Property System* (Kluwer Law International) and *Valuing Technology: Organisations, Culture and Change* (Routledge). Contact details: SATSU, University of York, York YO10 5DD, UK. Email: ajw25@york.ac.uk

Sally Wyatt is a Reader in the Department of Innovation Studies, University of East London. She is currently based in the Department of Social Science Informatics, University of Amsterdam (until late 2001). She is interested in the relationship between technological change and social exclusion and in the limits of constructivist approaches to analysing technological dynamics. Contact address: SWI, Universiteit van Amsterdam, Roetersstraat 15, 1018 WB Amsterdam, The Netherlands. Tel. +31 (0) 20 525 6765. E-mail: wyatt@swi.psy.uva.nl

Foreword

BY PROFESSOR BARBARA ADAM
University of Cardiff

Techno-scientific innovation creates unknowable futures. At the turn of the last century, for example, it was difficult if not impossible to imagine that a system of air transport would connect the cities of the world; that people could watch in their living rooms events taking place on the other side of the globe; that nuclear bombs would be dropped on Hiroshima and Nagasaki. True innovations are not predictable on the basis of past knowledge. Past experience, in other words, cannot serve as an indicator of the effects of scientific developments. It can only ever demonstrate the uncertainty of the future for societies committed to the industrial-scientific-technological way of life. A second clue to the uncertainty of techno-scientific futures relates to the scale of change brought about by those scientific technologies and to the speed and intensity of changes achieved. Globalised hazards, as well as accelerating cycles of innovation and obsolescence, create out-of-sync time frames. They produce mismatches between the time scales of invention, productivity and periods of waste and pollution, between benefits and hazards, between threats and ameliorative action, between contamination and visible effects, between effects and control.

Globalised techno-science of the early 20th Century dictates action in our present just as the techno-science of today creates future presents for our successors. Since these successors have no means of influencing societal decisions today it is the socio-political task of the present to find ways to encompass not just current socio-economic needs but also those of the future. This is clearly not an easy task, nor is it a task that many scientists and politicians are keen to engage in. Uncertainty is one important reason for its evasion.

Sociology has an extremely important role to play here in terms of explanation and rendering visible what are at present taken-for-granted and disattended social processes: how the future is created, constructed, contested, colonised and consumed; how it is materialised, managed and 'mastered'; how opportunities are created for some at the expense of others; how uncertainties, indeterminacies and contingencies are handled and how the profit potential of innovative technologies with unknowable side effects is played off against caution and precaution. The task for social science, therefore, is not to create ever better models to predict the future, identify trends and provide prognoses. Rather, it is to identify the field of contestation, show the intricate networks of interaction, establish the role of narratives, metaphors and 'grammars of the future', disclose power relations and their effects on alternative futures.

Contested Futures rises to this challenge. It is a timely venture into a sociological area of inquiry that will become a dominant focus for the social sciences as the new century progresses. *Contested Futures* analyses new problems and thus assists the social task of finding new solutions. Such sociological inquiry not only increases understanding but, more pertinently, it creates the necessary base upon which social responsibility for the future can be built and established.

Acknowledgements

A number of initiatives, projects and people are gratefully credited with having given this volume the future that it now enjoys. First, much of the thinking and discussion that takes place in the pages of *Contested Futures* stems from two research projects supported by both the European Commission (TSER-SOE1-CT97-1056) and the UK's Economic and Social Research Council (L323253035). Both projects (including teams from the UK, Spain and the Netherlands) have sought to explore the role and significance of 'the future' to organisational, scientific and technological change. We would particularly like to thank Barbara Adam for writing the foreword and those colleagues who accepted our invitation to discuss these issues at a workshop (*New Health Technologies and Future-Oriented Coordination*) held at the University of Twente, The Netherlands, September 1999. Thanks also to Lynn Kilgallon whose editorial skills hurried *Contested Futures*' future to the present.

Part One

Time, Temporality and the Social Construction of the Future

1 Introducing Contested Futures: From *Looking into* the Future to *Looking at* the Future

NIK BROWN, BRIAN RAPPERT AND ANDREW WEBSTER

The future is like a dead wall or a thick mist hiding all objects from our view (Hazlitt, 1822).

Like every thing future, all speculations on this subject must... be in a measure uncertain (Greenleaf, 1816).

Introduction

The *British Medical Association* recently published a book called *Clinical Futures* in which the stated aim of its editors was to 'redress a balance and create a forum' where the perspectives of clinicians rather than 'political, social, economic, legal, and organisational theory [might] take a freewheeling look at the likely trends in diagnosis and treatment over the coming decades... The intention is to bring the imaginative conjectures of clinical investigators to the fore of thinking about the future of health policy. We want to start a process that will strengthen the sometimes muted voice of physicians' (Marinker and Peckham, 1998).

Whether or not one agrees with the sentiment that 'the future' has been dominated by social science at the expense of clinical voices in health policy, there could be no better illustration of the way in which the future comes to be defined as a contested object of social and material action: if actors are to secure successfully for themselves a specific kind of future then they must engage in a range of rhetorical, organisational and material activities through which the future might be able to be 'colonised'. Whilst *Contested Futures* and *Clinical Futures* share a perspective on the future as a discourse in which different voices vie for ascendancy, they differ in one fundamental respect. In the very best of futurological tradition, *Clinical Futures* is a book about what the future should or might look like. *Contested Futures* does not postulate on the probability of one future against another nor generate normative prescriptions about particular futures. Instead, the intention here is to turn the analytical gaze towards the phenomenon of future orientation itself. The purpose of this analysis is not the future *per se*, but the 'real time' activities of actors utilising a range of differing resources with which to create 'direction' or convince others of 'what the future will bring'. As such, our purpose is to shift the discussion from *looking into* the future to *looking at* how the future as a temporal abstraction is constructed and managed, by whom and under what conditions.

This is done through exploring the contested future(s) of various sciences and technologies, not least because the experience and projections of late-modern society are, arguably unlike other periods, increasingly framed by techno-scientific language. Typically, our visions of the future are dominated by new technologies. The magazine *Newsweek* (1999/2000) recently asked what we can expect in the 21st century. The future outlined was one almost exclusively framed in terms of the implications of advances in science and technology: gene therapy and nanotechnology will cure disease, cars will drive themselves, pigs hearts will be used for organ transplants, computers will become an even more ubiquitous part of life, the Internet and the Cybercafe will become the venue of choice for our relationships, and so on. Little space is made for questions of human relations that are not structured around or presented as a consequence of the very latest gadgets. The optimism expressed for the future is coupled with a quality of inevitability in the development of particular technologies. While gene therapy might entail significant dangers, there is really no choice but to push ahead. The only question left is how fast particular advances will come.

In contrast to such treatments of the future, the contributors to this volume all share a concern to understand how it is that some futures come to prevail over others, why once seemingly certain futures happened to fail, how other futures are marginalised as a consequence of the dominant metaphors and motifs used in every day life, and the consequences of particular framings of the future.

There are a number of themes which have guided this project and which inform, in one way or another, the contributing chapters:

- Futures as Contested
- Agency, Action and the Future
- Path Dependencies, Determinance and Lock-In
- Contesting Future Models
- The Creation of Future Spaces/Moments - The Orchestration of Opportunity

It should be clear from this list (which is explored in detail below), that we do not see the future of science and technology as in any way the result of a linear or naturally evolving process. It is not simply that the future is always uncertain and so possible in multiple forms; this could quite clearly be a position adopted by someone who could still hold to an evolutionary perspective on science and technology (Basalla, 1988; Nelson and Winter, 1982). It is that the future of science and technology is actively created in the present through contested claims and counterclaims over its potential. Like all discourses, 'the future' is constituted through an unstable field of language, practice and materiality in which various disciplines, capacities and actors compete for the right to represent near and far term developments. By all measures, the future has become a big business. Witness in this regard the importance of the future in contemporary fiction and film, the need for public policy dealing with the environment to justify itself in terms of long term sustainability, the growing market for scenarios, Foresight, and horizon scanning in organisations, and even the emergence of book such as this one.

In many ways, the manufacturing of 'the future' is no different from discourses about 'the past' - not least, the history of science and technology itself. The history of science as recounted in most textbooks gives little idea of the contested futures that once shaped the development of what is considered to be the 'scientific canon' of today.[1] At the same time, it may well be true that those voices who sought to articulate what we might call 'past futures' were fewer in number and occupied positions of social status which meant that the 'contest' was narrower, more confined and less structured by liberal democratic discourse (Held, 1987). Today, the voices which declare a right to speak to the future are arguably more numerous, drawing on much wider, more heterogeneous forms of social authority. This is likely to mean that stabilising future expectations will become more and more difficult, especially for those charged with their social management, that is, government and regulatory agencies.

Futures Contested

As implied by the use of the plural noun 'futures' in the title to this volume, a common finding amongst its contributors is that futures occupy a contested terrain rather than being singularly or consensually defined. Social actors, at individual, institutional or wider cosmopolitan levels construct future expectations which may run in parallel with and contest each other, occupying different time-frames and carrying different interests. Therefore, we need to ask how and why futures are contested, and how future scripts are stabilised around a specific set of expectations and practices. How do actors seek to engage with and manage the promise and risks their futures are perceived and portrayed to hold?

The issues addressed in *Contested Futures* resonate with key tensions and ambivalences identified in contemporary commentary on Western society, such as the pervasiveness of discourses of risk, reflexivity and future indeterminacy. Beck, for instance, notes the almost crippling degree to which technoscience is widely seen to be both the source of terrible risks and yet the only plausible solution to determining impact and the deployment of ameliorative measures (Beck, 1992, 1995). Such strains are invariably refracted through a heightened public and political reflexivity evident in, for example, precautionary regulatory regimes. This relatively new arrival to what we might call 'future governance' betrays an understanding that sources of hazard require caution even when clear causal connections between today's actions and tomorrow's threats are vague or even improbable (Adam, 1998). So futures are not only contested in respect to a plural politics but also in respect to differing degrees of indeterminacy which have fragmented the basis on which we engage with the future. Hence, it is no longer the case that previous disciplinary approaches to the future have the value or legitimacy they once had and, instead, new voices have entered the discursive fray of the future's politics.

The future is also contested in respect to the relationships between acceleration, speed and an ever receding horizon action and agency. Anthony Giddens has contended that 'ours' is the most future oriented society there has ever been. That is to say it is the one most 'preoccupied' with the future, precisely because, unlike 'traditional' society it no longer has a sense of control over the future (Giddens, 1998). Saturated, as it is with competition, risk and knowledge intensity, planning has become more, not less, indeterminate:

> ... there is no way that the accumulation of knowledge will allow us simply to colonise the future, to carve out the future as a space which we can just invade and colonise. The very development of knowledge

actually makes the future more rather than less opaque... In the Middle Ages there was no concept of risk because it was not required. The idea of risk only develops when you have a society, which actively tries to break away from the past and conquer the future. In traditional cultures there is a sense in which there is no notion of the future... as a separate temporal domain because the future looks very much like the past. But with the rise of Modernity you have a society bent on changing the future. The Enlightenment philosophers saw the future as a territory you can occupy and colonise... and get away from God and dogma and from the influence of the past to shape the influence of the future (Giddens, 1999).

For some, such as Baudrillard (1994) this loss of control over the future is as much to do with our collapsing sense of control over the past, where the meanings and 'truths' of a linear history no longer make any sense. It may also be seen as a result of what Giddens (1998) calls the 'end of nature' where we no longer are concerned by what nature can do to us, more what we have done to nature.

In respect to the social theory of temporality, these considerations have forced on us an entirely new conceptual means of engaging with the future. Theorists like Adam have pointed to the helpful merits of recent science discourses ranging from eco-evolutionary theory and chaos theory in thinking about our relationship to the future. For example, evolutionary theory has drawn our attention to the long historical processes of species development now potentially threatened by the compressed time frames of genotechnology. The second law of thermodynamics has whittled away our sense that we can reverse today's actions in the future by stating that rearrangements of energy are inherently unidirectional involving irreversible states of ageing and decay. Where Newtonian physics once propagated the view of an equally proportionate relationship between causes and effects, chaos theory has unseated this equilibriumism with an appreciation of how small changes disproportionately reverberate throughout entire systems in nonlinear ways. Such is the complexity of the relationships between processes which operate to different temporal principles that 'identical paths and outcomes would constitute a miracle' (Adam, 1998 p. 46).

Deconstructing past agendas and opening up new ones generates uncertainties as well as opportunities. For those in industry this has meant a growth of markets for new (both process and product-based) technologies which are increasingly knowledge-intensive - such as ICTs (Information and Communication Technologies), genetic diagnostics or the rapidly expanding digital electronics sector (Gibbons, 1994). Yet these markets are not to be constructed simply through the 'logics' of competitive advantage: they are hedged about by technical, cultural, economic and regulatory constraints that make them highly unstable. It is, therefore, perhaps not surprising that the competing innovation agendas being pursued today are, to some degree, marshalled and corralled by the policy sector through future-oriented government programmes such as Foresight that has swept through the world's most industrially active nation-states over the past decade or so (Gavigan, 1997; Martin, 1995). Indeed, the emphasis on the need to create a sense of shared future through policy instruments like Foresight programmes are indicative of the need to address and manage the future's fragmentary or indeterminate character. When the future can no longer be expected to follow on neatly from the past, then imaginative means must be employed.

The analysis of articulated futures is crucial because it is in the shared times of the new millennium that contested futures in science and technology are most apparent and where the future agendas of distinct constituencies have to take account of one another. The chapters in *Contested Futures* illustrate the different ways conflicts are expressed and the strategies different actors use to embed and secure a particular reading of the future.

Agency, Action and the Future

A particularly thorny problem for those who seek to understand the way in which future technologies and knowledge are secured is where to locate agency. How is it that things come to be the way they are and, more importantly, as a consequence of whose actions, under what conditions, and to what effect? In some circumstances futures are formulated for technologies by concentrating agency in a product or even a scientific claim, thus obscuring the many other actors and contingencies upon which that product's future once depended (Deuten and Rip, Chapter Four). This may be especially the case, for example, when scientists claim that they have made a dramatic 'breakthrough' which opens up new work: breakthroughs are often presented as speaking for the future implications of a technology when in fact the future career of the specific knowledge-claim is still yet to be built (Brown, Chapter Five). Sometimes, rather intriguingly, scientific claims to novelty can bite back and create unanticipated problems for scientists: this happens for example when patent claims filed on novel inventions for future development are refused because the patent examiner cites prior work of the *same* (and not only other) scientists, on the grounds that this earlier work anticipated future developments, including that lodged by the patent claim (Webster and Packer, 1996).

Agency is also often attributed to a particular technology itself, seen as unfolding or developing naturally along identifiable lines as its 'self-evident' benefits are taken up by users. This view is typically found in commentaries on the 'future impact' of a list of new technologies. Here, the key question is not *whether* a certain option will be pursued, but rather *when* it will come in being. In its more elaborate forms, such thinking has led to some of the more sweeping claims about the dynamics of technology in industrial society (Kerr, 1960; Bell, 1974). Technology here works to promote a convergence in social structure by virtue of its organising potential, especially in the workplace. In such a situation, human agency becomes reduced to engaging in behaviour to ensure the speedy uptake of particular technological possibilities. The force implied in this attribution of agency is that one can either ride the wave of advancement or drown in the waves of progress.

Powerful political narratives - such as those associated with both communism and fascism - also capture future promises. On the one hand, communism implies a future where the perfect society is to be secured through action, while fascism is typically motivated on the basis of the perfect society already being here (in the highly charged language of dominant essentialist ethnic identities, such as the 'Aryan race'). The 'liberal democratic' political narratives pervasive in many Western countries are, of course, not without their commitments. Those in subordinate positions are promised the rewards experienced by the well-off if they play the rules of the game set out and reinforced by those in dominate positions. There is too a politics of the future which involves different forms of selection and boundary work - that of gender and the patriarchal future which informs literature, including science literature. This issue is touched on in this book through an examination of the gendering of science fiction. Temporality - past, present and future - can, whether through a language of ethnicity or gender, be captured and be enacted through the articulation of both forms of discourse.

Just as commonly, agency is projected into the future itself. Examples of this kind are implied in the grammar of the future through which actors speculate upon 'what the future will bring' and near-certain expectations of what the future 'will be like'. Here time itself is reified into acting determinately raising broader questions of how it is that agency becomes a property of time and 'the future' rather than the ordering practices that produce it. Such practices thus become concealed behind the future acting as a powerful agent on our present.

The question arises then of how metaphors, agendas, scripts, narratives, expectations and promises function as methods by which actors seek (either consciously or otherwise) to secure a certain kind of future for themselves and the many others who will be essential to a desired future. The issue of agency in the mobilisation of the future provokes a number of questions: How might we understand the relationship between expectations and their impact on behaviour or the outcome of events? Where should the analyst locate future-oriented agency, in a future representation, in its authors or in agents who cannot properly be said to harbour expectations of any kind at all? What assumptions about agency are embedded in the way we conceptualise temporality and the future? What analytical and methodological approaches are available to think about agents acting in and on time, deploying representations of time and themselves becoming implicated in wider futures? How are expectations reflected in an actor's capacity to exercise future-oriented agency?

Path Dependencies, Determinance and Lock-In

While the deterministic implications associated with technological and scientific trajectories have been critiqued in science and technology studies (STS) and other fields, both within and outside of academic circles reference is often made to clear and persistent patterns of development. The trends toward the increased calculation speed of computers ('Moore's Law') or the enhanced transportation mobility of individuals are just two examples.

Economists, among others, utilise the notions of path dependencies, lock-ins and lock-outs in discussing technological change. Arthur's (1994) modelling of path-dependent 'lock-ins', that is, when technologies demonstrate increasing returns to adoption, is a good example here. The more technologies are adopted and their problems resolved, the better their performance and the greater their adoption. This clearly generates a powerful path-dependency over time, one which marginalises competing or new technologies. There is, however, nothing intrinsic to the technologies that makes one more superior to another. The difference lies not in the technology necessarily but rather in the contingencies of socio-technical circumstances and the play of institutional interests that favour one technology over another. A war-time crystal radio is as good at being a crystal radio today as it was then. But locate it in the wider heterogeneous context of contemporary expectations of digitised communications and new interfaces and it becomes easy to mistakenly think of the crystal radio as itself an 'inherently' flawed artefact. These processes can be seen to be at work in the adoption of medical technologies - such as a path-dependency that appears to be emerging around specific storage and analytical instrumentation used in genomics and bioinformatics-based research. Much of this is to do with standardisation and globalisation of a technology, but is also strongly associated with powerful institutional, bureaucratic and professional/expert interests: in some cases it would appear that a suboptimal medical technology system has been sustained for a number of generations as a result (Blume, 1992).

More substantively, over the past decade or so, there have been some powerful commercial, policy-related and intellectual narratives associated with the anticipated future impact of three scientific fields: biotechnology, new materials, and ICTs. These three have often been characterised as generic 'enabling' technologies whose presence is evidenced across a wide number of traditionally discrete product sectors. This encourages a policy lock-in around each of the three areas and a wider cultural narrative which requires public and private sector actors to attend to the demands and opportunities they are said to bring. Organisational narratives, for example, are increasingly framed by the 'need' to secure the 'latest' IT systems in order to remain competitive and to manage information and human resources (McLaughlin et al., 1999). Corporations may build complete organisational strategy around such technologies (such as Monsanto did around certain forms of biotechnology in the late 1980s) and draft strategic 'road-maps' which anticipate future-oriented paths down which the firm will go.

What relevance then does the notion of 'trajectory' and related concepts of lock-in have for study of expectations and promises? In what ways are scientific and technological 'paths' deployed to justify certain actions, for example, by serving as a measure for assessing activities or acting as a self-fulfilling prophecy? How are trajectories maintained and contested? In what manner do they constrain future options? How do actors shift between attributing 'technical' and 'social' factors with a determining force (see MacKenzie and Wacjman, 1998)?

Contesting Future Models

It is often acknowledged that the uncertainties surrounding scientific and technological developments and the general inadequacies of forecasting, make attempts at modelling the future acutely problematic. How in such circumstances do actors attempt to model future developments and what functions do such models serve?

Models serve different purposes. Some, such as the global environmental models that informed the now classical text *Limits to Growth - a Blueprint for Survival* (Meadows et al., 1972) portray the future as highly knowable - that behavioural change is needed if we are to secure our future. Uncertainties are replaced by strong probabilities here and there is little sense of such models offering a range of options which might meet the same ends. This type of model is much less in vogue today, and instead we have a range of futures languages that structure uncertainty and multiple choices into the model itself. Models become more provisional, used as aids to think through possible futures which might then be actively pursued. Scenario analysis (see e.g., Ling, 1998), horizon scanning, future visioning (O'Brien and Meadows, 1998), 'back-casting' (Inayatullah, 1995) are all deployed to help manage the perceived complexity of things to come, a complexity that is regarded not with trepidation but embraced as something to be engendered by an engagement with the future: as was observed at one recent 'futures' event, 'forward planning is how to make boundaries more fluid' (Kings Fund, 1997). In doing so, perhaps unintentionally, scenarios provide a means of distancing oneself from present arrangements and thus in some circumstances enabling a space for criticism. Conventional approaches to future planning based on forecasting requirements have been jettisoned for many purposes because the future is seen to be 'more unpredictable' than before as a consequence of the pace of innovative change, the heterogeneity of knowledge and a widening of relevant constituencies (*Future Studies*, 1998).

It is not clear how the 'pace' of future change is measurable, whether and to what extent we can intervene in the accelerating tempo of industrial change. The language of future choice tends to argue against the notion of change as being deeply structural, by suggesting that there is always everything to 'play for'. This is embedded in the recently fashionable notion of 'stakeholder' which suggests that futures are open to different investments made by distinct social groups. The arrival of a stakeholders' future(s) rather than that of a social class or similarly broad social collective works against the view that constructed futures are also about constructing or realising collective goals. The future seems no longer to be produced collectively through some subscription to a wider collective set of norms, but consumed through disaggregated stakeholder populations.

The issue of modelling the future is one which needs to consider not only the assumptions of the modeller but also the modelled - that is, do groups actively co-construct their visions of the future, or are they simply co-opted into a particular text which for various reasons commands acceptance? When does the need to make decisions today foreclose future alternatives and who should be involved in attempts to derive future scenarios? This leads us to our final broad theme shaping the contributions of this book.

The Creation of Future Spaces/Moments - The Orchestration of Opportunity

This book provides an opportunity to explore the closing off of futures as well as the challenges made to exclusionary futures. In what way do future scripts carry assumptions and presumptions which express forms of social inclusion and exclusion inasmuch as they typically mark out the boundaries of the future which people must occupy? In what way does 'the future' carve out and make available, therefore, different temporal 'spaces' or 'moments' that constituencies of interest occupy in respect to occupation, expertise, gender, age? How should we regard such attempts? This clearly might be regarded in positive or negative terms - as in political agendas which presume on the one hand a future based upon an inclusive democracy, and on the other, an exclusionary set of interests (as in apartheid). To what extent are alternative futures able to enter into existing discussion about the future?

The structuring of an individual's social future is apparent through the patterns of social mobility which are structured not randomised, and which allocate more, or less, advantageous life chances for different social groups. Science and technology might well add to this process through the structuring of life chances via the mergence of a new social eugenics which some believe will begin to generate new types of disadvantaged futures for those carrying specific genes. The geneticisation of social stratification systems has taken place before, of course, through both ethnic discrimination and the horrors of ethnic 'cleansing'.

Elsewhere, powerful discourses can prevent or marginalise alternative futures, such as those envisaged by radical science, by alternative science and technology, by deep ecology, and so on. It is true that orthodoxy has had to respond to the challenge posed by these alternative voices, but typically it has done so either by rejecting them out of hand or co-opting them when this seemed a more appropriate strategy. This can be quite variable over time, as the uneven fortunes of alternative medicine over the past 200 years demonstrate. Co-option of alternative future narratives, also normally involves a translation of those narratives into the language (and future) of dominant voices: so 'acupuncture' has become 'transcutaneious nerve stimulation' or having a positive outlook during illness is translated into the indices of psychometric 'hope scales' and so on.

The orchestration of futures by powerful groups suggests that, perhaps, the future is not so uncertain and unpredictable as some of the other themes described above. It is as important then to recognise the need to strike a balance in our analysis between interrogating how futures are constructed and the openness and contingency this involves, and who constructs the future, and the relative closure that this involves.

Having presented some of the general themes to which *Contested Futures* will respond, we now turn in more detail to the ways in which these themes inform the contributions of different chapters.

Structure of the Book

Part One: Time, Temporality and the Social Construction of the Future

The performative nature of accounts of the future is an ongoing theme in each of the chapters in this collection. Our examination of technoscience in this way begins with Chapter Two by Mike Michael who sketches out the anatomy of that performativity by identifying key parameters by which the future is constructed. Parameters such as the temporal distance of future events, the pace of the passage of time, and our conceptions of experience, give representations of time and the future a varied functionality. The central concern of Chapter Two is to consider how the specific content of a future implies particular subject positions, identities, relations of power, and versions of community. As we will see in later chapters, the parameters described by Michael underlie a wide range of accounts of science and technology. But representations of the future are not amaterial. This chapter also locates futures in material world and considers how this materiality shapes both representations of 'the future' as well as the 'present' in which those futures are made to perform.

Part Two: Language and the Social Rhetoric of Technical Futures

This section follows from the previous discussion on performativity by asking how it is that language and materiality are both, within given limitations, mutually constitutive of various futures. With this in mind, each of the contributors has sought to arrive at what we might call a symmetrical sociology of the future, by exploring the connections and dissonances between the materiality of technoscience and its articulation in language (narratives, metaphors, scripts, promises, idioms and icons).

Harro van Lente, in Chapter Three, examines the way that promises and expectations are able, or not, to oblige certain arrangements of socio-materiality into future place. These prescriptions can be as pervasive as the Western discourse of 'technological progress' itself or as micro-political as promises. Promises, Van Lente argues, have a certain temporal inertia in that they can quickly mature into harder requirements, contracts, mutual obligations and eventually path dependencies. The chapter discusses a range of empirical cases including media technologies and the Human Genome Project.

Future stories are not just a way of persuading certain futures into being, they can also serve as metaphors for the innovation process itself. Deuten and Rip (Chapter Four) use narrative and story as a means of understanding how actors oblige one another into conformity with certain sequences of story telling and role playing. Narratives usually have a certain plot, at the beginning of which a 'stage is set', roles are assigned to the key actors who will be instrumental in resolving certain basic narrative problems and tensions. It is a shared familiarity with the narrative form, in the context of a biotechnology firm in this case, that enables us to recognise 'where things are going' and how elements in the story should, *in the future*, behave. Narrative is then constitutive of the future.

Narrative is only one way of expressing the future oriented character of the innovation process. In Chapter Five, Brown examines the way the breakthrough motif has only recently come to populate the public cultural representation of science and technology. Like narratives, breakthroughs imply a certain normative story structure beginning with a problematised present, an impasse to which a development may come to count as a breaching. Accordingly, the breakthrough motif, now probably the most ubiquitous motif of progress in popular science (especially in medicine), serves to structure both the past and the future. By way of illustration, the chapter uses recent case of Dolly the cloned ewe and research into animal-to-human transplantation to argue that breakthrough invariably misrepresents the actual doing of science and over-idealises what most developments are able to deliver.

Sally Wyatt's discussion (Chapter Six), focuses explicitly on the formal properties of metaphor, and illustrates the powerful future-oriented metaphors that have helped shape the future of Internet technologies. Metaphors can imply a certain temporal and spatial directionality. For example, the much vaunted 'information superhighway' conjures up images not only of a synchronic spatial infrastructure for the rapid movement of binary ones and zeros, it is also the techno-metaphorical 'conduit' to the Internet future itself. Like narratives, metaphors do not hover outside techno-scientific institutional arrangements, rather they are constitutive of those arrangements and thus require constant analytical vigilance to understand their implications.

Part Three: Past Futures

Of course, concern for the future of science and technology is not solely a contemporary phenomenon. 'Past futures' provide fertile ground for illustrating how space was made for alternative worlds, what functions 'visions of tomorrow' have served in the past, and how certain options became marginalised. The two chapters in the third section of this book examine a number of past futures. As is clear in both, the stories told about technoscience futures owe much to the time and place of the actors who formulate them. Questions about whether the future is seen as one which engenders feelings of pessimism or optimism and whether it is one where technology acts to co-ordinate human activity or serves as a facilitator for individual autonomy, cannot be divorced from the situated lives of those contemplating them.

Hilary Rose begins this section by discussing the visions and the commitments present in science fiction. As a genre, science fiction has long offered a means for writers to explore alternative worlds free from many of the limitations of other forms of fiction. While much of the mainstream portion of this literature has eschewed from reflecting on social relations or been limited to considering the implications of impending scientific and technological change for society, Rose identifies two periods of (English language) science fiction which provided insight into possible technological and also social *transformations*: the time between the world wars and the period after the women's liberation movement in many Western democracies. Science fiction offers an opportunity for the critical evaluation of social relations. The writing associated with thirties Western Marxist and second wave feminist critiques took impetus from and inspired social, political, and economic reform. Feminist science fiction, in particular, has demonstrated that a particular form of technoscience is not necessarily desirable and certainly not inevitable; the future is socially constructed and, as such, is continuously open to reconstruction.

In Chapter Eight, Geels and Smit examine why particular expectations about the future fail to materialise. As they illustrate in the case of the implications of information and communication technology for traffic and transportation, images of the future are bound by existing practices. Although many of us are gripped by the possibilities offered by forthcoming technology, our expectations for the future are typically extrapolations of the present. Innovations are framed in terms of allowing us to do things faster, over a greater distance, and more conveniently than they, are done today. In doing so, however, crucial issues are obscured, especially those related to importance of the social embedding of technology for its effects. This can be witnessed today in the continuing enthusiasm for teleworking as an alternative form of work. Despite the considerable public and government attention to teleworking, so far its spread has been fairly limited and arguably has led to opposing practices than those it was supposed to encourage. Many of the reasons for this derive from the failure to acknowledge the inter-relation of the technical and the social. In addition though, the analysis of Geels and Smit alert us against the all too easy approach of simply criticising the short-sightedness of scientists, engineers, or policy makers. Rather, following the work of Harro van Lente, expectations of the future must be seen in terms of their performative character. They are strategic resources in political and technological agenda-setting processes.

Part Four: Future Science, Future Policy and the Management of Uncertainty

Part Four of *Contested Futures* turns our attention to the policy domain, and the various ways in which the discourse and practice of a future-oriented agenda is mobilised at the level of the organisation, a particular technology sector, a policy domain (such as health) or at a national level. Much of the analysis focuses on the dynamics through which formal government programmes such as Foresight are articulated and given very different expression in distinct socio-technical contexts. Each of the four chapters explores the role of expectations, agendas, and the management of future uncertainties surrounding science-based innovation.

In Chapter Nine, de Laat shows how, as he says 'every technical choice implicitly involves a hypothesis on how, *socio-technically*, the future may be organised'. That is, the emergence of a new technology always depends on those developing it making assumptions about its future location in a wider technological ecology. Such assumptions are, says de Laat, built into 'fictive scripts' which once deployed bring greater momentum to the innovation process. One of the more significant of these scripts is that of Foresight, which the author unpacks and critiques especially when conceived of as a formal methodology for 'forecasting' or predicting technological futures. Instead, Foresight acts performatively - a notion seen already in Part One and Two of this book - to construct a future, and does so through socio-technical networks.

The focus on networks provides a key framing for the chapter by Nelis, which moves the discussion on to provide a very detailed consideration of future-oriented co-ordination among the different networks found in UK and Dutch health innovation and delivery in the field of genetic diagnostics. Much has been made of the 'impact' of the new genetics on health care, and in particular through the diagnostic and therapeutic capabilities genetic research is supposed to bring. Nelis shows, however, that the character and significance of genetic diagnostics will depend crucially on the institutional 'configuration' within which it is expressed. By showing how a number of socio-technical, organisational and cultural uncertainties are managed by different constituencies and their networks in the two countries, she shows how Foresight can be perceived as more or less valuable as a vehicle through which configurational relations might be changed. Here again, there is a sense of the need to attend to the varying potential performativity of Foresight.

While Nelis draws our attention to the management of expectations and uncertainties, Sanz-Menéndez and Cabello argue for a more cautious approach towards the importance of future expectations on technology shaping. They provide a detailed examination of the meaning of expectations for actors engaged in innovation - especially within the firm - and argue that our analytical attention needs to focus as much on the routinised practices and repertoires carried from the past into the present, for it is through these that actors make most decisions and choices seen as 'appropriate' to the circumstances they confront. A rather different emphasis can be given on the learning practices that innovation actors deploy in shaping strategic decisions oriented towards the future. Here then, as Michael argues in Chapter One, constructions of past practice can be as powerful a register for future (and present) behaviour as those explicitly tied to a rhetoric of future change.

Finally, Chapter Twelve concludes the book with Ling's careful exploration of the British National Health Service (NHS) and the changing fortunes of professional expertise and its entrusted status as deliverer of health goods, now and in the future. He shows how NHS policy actors have had to struggle with a loss of both normative and technical control over the future agenda for health care as new demands, new entrants to the health market, and newly 'empowered' patients deconstruct the stable NHS world they once enjoyed. He illustrates his analysis of this more fractured and fragmentary world through a consideration of the implications of genetics and informatics for the NHS. The new uncertainties they are perceived to have brought can no longer be managed by either a patrician or a bureaucratic elite, but through a relocation of responsibility onto the shoulders of the citizen accompanied by an embracing of more reflexively-based tools for managing the new uncertainties, such as scenario analysis. These developments, however, need to be located within a close analysis of how agendas are set, by whom, and for whom. The future, in this sense, is as political a performer as it has ever been.

Note

1. For exemplary exceptions to this see Miller (1988) and Hughes (1983).

References

Adam, B. (1998) *Timescapes of Modernity: The Environment and Invisible Hazard*, Routledge, London.
Arthur, W.B. (1994) *Increasing Returns and Path Dependence in the Economy*, University of Michigan Press, Ann Arbor.
Basalla, G. (1988) *The Evolution of Technology*, Cambridge University Press, Cambridge.
Baudrillard, J. (1994) *The Illusion of the End*, Polity Press, Cambridge.
Beck, U. (1992) *Risk Society: Towards a New Modernity,* Sage, London.
Beck, U. (1995) *Ecological Enlightenment: Essays on the Politics of the Risk Society,* Humanities Press, New Jersey.
Bell, D. (1974) *The Coming of Post-Industrial Society*, Basic Books, New York.
Blume, S. (1992) *Industry and Insight*, MIT Press, Cambridge, MA.
Future Studies (1998) *The Metropolitan*, Issue October 30.
Gavigan, J. et al. (1997) *Overview of Recent European and Non-European National Technology Foresight Studies*, Report No. TR97/02, CEC: IPTS Seville.
Gibbons, M. (ed.) (1994) *New Production of Knowledge: Dynamics of Science and Research in Contemporary Societies*, Sage, London.
Giddens, A. (1998) Risk Society: the Context of British Politics, in J. Franklin (ed.), *The Politics of Risk Society,* Polity Press, Cambridge.
Giddens, A. (1999) BBC Reith Lectures – Runaway World. View online at: http://www.lse.ac.uk/Giddens/lectures.htm
Greenleaf, M. (1816) *Distr. Maine*, 136.
Hazlitt, W. (1822) *Table Talk, Essays on Men and Manners,* Bell Bohn's, London.
Held, D. (1987) *Models of Democracy.* Polity Press, Cambridge.
Hughes, T.P. (1983) *Networks of Power: Electrification in Western Society, 1880-1939*, Johns Hopkins University Press, Baltimore.
Inhayatullah, S. (1995) Future visions for S E Asia: some early warning signals, *Futures*, 27, 6, 681-688.
Kerr, C. (1960) *Industrialism and Industrial Man*, Basic Books, New York.
Kings Fund (1997) *Report on Kings Fund Symposium on Health Futures,* London, November 10-11.
Ling, T. (1998) *The Madingley Scenarios*, NHS Confederation, London.

MacKenzie, D. and Wacjman, J. (eds.) (1998) *Social Shaping of Technology*, 2nd Ed. Open University, Milton Keynes.

Marinker M. and Peckham M, (eds.) (1988) *Clinical Futures*, BMJ Books, London.

Martin, B. (1995) Foresight in Science and technology, *Technology Analysis and Strategic Management*, 7, 139-168.

McLaughlin, J. et al., (1999) *Valuing Technology*, Routledge, London.

Meadows, D.H. et al. (1972) *The Limits to Growth*, New American Library, New York.

Miller, C. (1988) *When Old Technologies were New*, Oxford University Press, Oxford.

Nelson, R.R. and Winter, S.G. (1982) *An Evolutionary Theory of Economic Change*, Belknap Press of Havard University Press, Cambridge, MA.

Newsweek (1999-2000) *Newsweek: The International Newsmagazine*, December 27-January 3.

O'Brien, F.A. and Meadows, M. (1998) Future Visioning: A case study of a scenario-based approach, in Dyson and O'Brien (eds.) *Strategic Development*.

Webster, A. and Packer K. (1996) Patenting culture in science: re-inventing the wheel of scientific credibility, *Science, Technology and Human Values*, 21, 427-453.

2 Futures of the Present: From Performativity to Prehension

MIKE MICHAEL

Introduction

What is the future? This simple question masks a complex of issues. For example, do we mean by this the substantive, content-ful future - what specifically will happen in time, subsequent to where we find ourselves now? Or, do we mean the formal future - what is meant by the 'future' experientially? As Adam (1990) has shown in exploring the possible roles that time and the temporal might play in social theory, time (let alone the future) is difficult to pin down. Time can be seen to be passing, a stream in which we are caught, periodised into our familiar units of seconds, minutes, hours and so on and so forth. But this is one particular and peculiar version of time. It is what Adam calls 'clock time' - spatialised, abstracted, mathematicised - and it is linked, in part, to industrialisation. It is a time of its time, so to speak. Now, this sort of representation of time certainly structures our experiences not least in that it resources, for example, our biographical narratives (see Freeman, 1993). However, it can be contrasted to being 'in time' - or 'lived temporal duration' as Adam, glossing Bergson, puts it (1990 p.4). According to this latter view, the past, the present and the future are not situated on a line, that is to say, linearly spatialised. Rather, 'the past and the future... are constantly created and recreated in a present' (p.24). Thus Adam, glossing Mead, tells us the 'locus of reality is the present' (p.39) - past and future are not beyond the present but representations within it. The 'real past, just like the real future, is unobtainable for us, but through mind is open to us in the present' (p.39). Put rather too simply, we are only in the present, and in 'managing' in the present we deploy representations of the past and the future.

In the context of the concerns of this volume, to 'manage' can take on a number of forms which, for present purposes, can be reduced to two basic ones. On the one hand, there is the 'production' of some future state of affairs through the application or development of scientific knowledge or technological artifacts. On the other hand, there is the organisation of some future state of affairs in which science and technology might 'flourish'. To enunciate a research question, to formulate a research programme, to outline a prospective technological system, to posit the coordination of industry, government and university sectors in the pursuit of 'sound technoscience' - all these entail statements made (or rather performed) in the present that draw (on) the past and the future. That is to say, there is a 'fabrication' of past and future that make these enunciations, formulations, outlines and positings seem eminently sensible and do-able. Thus, the past is represented as entailing some problem (e.g. the chaotic state of science policy), some absence (e.g. the lack of transplantable human organs), some wrong (e.g. environmental degradation), and the future is represented as the 'place' where solutions are realised, presences manifested, and wrongs righted. This is the formal 'lived temporal duration' model of time sketched out above: the present is the locus (which we can never leave) in which are drawn together substantive representations of particular sorts of 'sociotechnical' past and future.

As such, when we speak of the future, we should say we are dealing with substantive representations of the future. Part of the task of this chapter will be to survey, albeit sketchily, a number of the ways in which the future is represented. Now as we have noted, for Mead the 'present' which is inhabited by these representations is that of the mind. But such representations are also relational insofar as they are 'between' persons, instrumental in the structuring of relations. As such, and in keeping with recent accounts of representation (e.g. Potter and Wetherell, 1987; Law, 1994), we should look at these representations in terms of performativity. In other words, we must ask how are such representations of the future so constructed that they 'perform' in such and such a way. Such performance can be said to entail a number of dimensions: the presentation of self, the production of subject positions for readers/viewers, the enrolment and alignment of various others, the bringing into being of a particular state of affairs. In surveying the contents of these substantive representations of the future then, I will attempt to draft a number of parameters by which the future is textually constructed. Part and parcel of this explication will be an exploration of the sorts of performativity associated with these futures. How do the specific content of a future imply particular subject positions, identities, relations of power, versions of community and so on? How are these invitingly portrayed? In pursuing this analysis, I draw upon the rhetorical analytics of Michael Billig (1987, 1991; Billig et al. 1988).

However, I also address another set of issues. I am interested in thinking about how the subject positions, identities, relations of power, versions of community and so on that are performed by representations of the future, are also reflected in (and mediated by) the materialised form of these representations. That is to say, these representations are not a-material, ethereal bits of floating culture: they are substantiated, objictified on paper, verbally, on the screen, pictorially (in this they do not differ from any other representation). The mode of travel of these representations, that is the medium through which they are communicated and the material routes they follow, affect the sort of future that is portrayed, and the sort of performance that is possible by a particular represented future. In the final section, I will explore the implications of thinking about representations of the future in this way, especially in terms of contested representations of the future of scientific and technological developments.

Surveying the Future

What follows is a tentative and thoroughly equivocal 'parametization' of representations of the future. That is to say, I want to explore some of the characteristics that representations of the future are likely to possess. For example, we might say that a future is represented as positive or negative, or as far from the present or relatively near-by. Clearly it would be foolish to claim an exhaustive account of the aspects by which the future is represented, not least because of the way that other cultures would dispute the very frame of the 'future'. However, we might say that within certain sections of western culture accounts of the future can be set out in terms of a number of parameters. Although there is a certain temptation to rank these parameters in terms of their basic-ness, I have resisted this. This is largely because of a lingering suspicion that stating that X is 'more basic' than Y (say 'distance from the present' is more basic than 'forms of rationality') might disguise the fact that X and Y are not separate variables (say, long-distance futures might be more prone to framing in terms of substantive as opposed to instrumental or bureaucratic reason). As such the list that follows should not be seen as hierarchical, its items should not be regarded as mutually exclusive, and each parameter should be assumed to be a conflation of other categories that would benefit from conceptual refinement.

The way I present these parameters is as dichotomies (some of which are really dimensions, as we might expect) such as near/far, good/bad and so on. However, in setting up the following dichotomies I do not want to make any claims about their actuality in an ontological sense. Rather, I want to focus on how these dichotomies perform. What does saying a future is near, as opposed to far, do for the sayer? Conversely, what are the sorts of readings enabled by saying that a future is far as opposed to close by? Indeed, my interest is in how these aspects of the representation of the future function rhetorically. As such, I draw, in particular, on Michael Billig's pioneering work on rhetoric, argumentation and ideological dilemmas. This ranks as one of the major contributions to recent social psychology (e.g. Billig, 1987, 1991; Billig et al. 1988) and is gaining increasing influence within the social sciences more generally. Here, I will give only a very brief overview of his approach in order to contextualize the later discussion. For Billig, ideology is not uniform, it is *dilemmatic* - it contains contrary themes and people's thinking reflects this. Accordingly, in order to understand the meanings and functions of people's utterances and writings, it is necessary to reflect upon what it is they are arguing against. Billig phrases this point in the following way: 'The context of opinion-giving is a context of argumentation. Opinions are offered where there are counter-opinions. The argument "for" a position is always an argument "against" a counter-position. Thus the meaning of an opinion is dependent upon the opinions which it is countering...' (Billig, 1991 p.17). This applies to the analysis of texts as well: 'Unless readers of propaganda understand the counter-position, against which the message is directed, they may be misled by the message' (Billig, 1991 p.19). As we shall see, when someone argues that a future is 'distant', there is a tacit argument against it being nearby. What is such an argument performing rhetorically? How is it occluding alternative representations of the future? How does it counter different arguments? How does it attempt to silence contrary voices?

   In what follows, then, I will provide a brief description of parameters, treating each as a dilemma (or dichotomy) in order to draw out a few observations as to how it might operate rhetorically, that is, performatively.

Distance: Distal versus Proximal

Unsurprisingly, any representation of the future will specify, in some degree, how distant it is from the present. Is it the future of a few weeks' time, of a few years' time, of a few centuries' time (and where and what is its scope)? Here, clearly we see the evidence of the operation of clock time and linear spatialisation. Before going further we should note that the unit of time used to measure this 'distance to the future' can itself have rhetorical effect. Thus what are the respective impacts of measuring a future's distance from the present in terms of weeks, in terms of parliamentary sessions, in terms of a technological age? Each of these units suggests a 'do-ability' - it is easier to accomplish something within a lifetime of a parliament than within an era of western civilization.

Be that as it may, the main point at issue is the way that distance is used in making a representation of the future 'perform'. A future state of affairs, say a certain degraded condition of the environment, can, depending on its distance from the present, signify very different things. Long distance would diffuse urgency: there is no dire need to do anything immediately, and in the meanwhile 'we' can wait for technology to develop fixes. Contrariwise, if this future is situated close by, then action is needed immediately: policies have to be developed and implemented, alliances and allegiances must be put in place, and so on. The rhetorical dilemmatic dimension of this parameter is fairly obvious. Great distance of a future *bad* facilitates a more 'relaxed' set of responses; shorter distance suggests more energetic efforts. The obverse applies to a future *good*. In the context of these more or less distant futures as well as claims and counter-claims about action, policy making and implementation can be done. A future represented as far distant can be used to warrant slowness of action, but it can also draw the charge that it serves in a tactic of delay. A near future can warrant swift action, but it can also attract the accusation that it is no more than opportunism on the part of the actor who gains from some sort of 'scare' or other. For example, biotechnological interventions might be seen as an improvement on changes wrought by evolutionary processes or traditional breeding programmes insofar as possible future genomes are represented as situated closer to the present. For supporters of biotechnology this is good because, in this context, the far distance of the products of evolution and traditional breeding methods signals inefficiency. In contrast, for detractors this is bad because the supposed safeguards built into evolutionary processes are by-passed - distance is what ensures some measure of safety. However, the twist here is that both these assessments of biotechnology rest on other futures: for supporters of biotechnology, there is a close-by risky future of starvation; for critics, there is a close-by future of viable ecological alternatives to biotechnology. Moreover, there are pasts that are recruited in the representation of more or less distant futures. Thus as Plein (1991) notes, in the promotion of biotechnology actors draw equations between past 'traditional' techniques and modern biotechnological ones. Such equations (or such differentiations) can serve in bringing the future closer - it makes it less distant, and thus less alien. As · we might expect, it is the absolute novelty of biotechnology that is argued for by its critics.

In sum, the distance of the future rhetorically positions the reader in a number of ways. It can urge immediate action or measured consideration, it can be used to accuse actors of tactical procrastination or opportunistic band-wagoning, and it can serve in the judgement of relative efficiency or risk.

'Subject': Individual versus Collective

By subject, I mean to suggest the entity that 'experiences' the future. This 'subject' may range from a human individual to a heterogeneous collective. Thus an individual sufferer may be seen as the bearer of future circumstance, or a generation may be seen to reap the (dis)benefits of a future state of affairs, or an ecosystem may undergo the future. Each of these subjects has different rhetorical functions. For example, Brown (1998) has shown how the biographies of individual human sufferers of cardiovascular disease can serve as powerful tools in the advocacy of xenotransplantation. Tapping into powerful narrative conventions, the biographies of sufferers that project into the future the inevitable death of the sufferer by virtue of the lack of human hearts, serve in setting up xenotransplantation as the only 'realistic' option. Of course, as many of us have experienced, this sort of narrative applies no less effectively to individuals or families of 'charismatic megafauna' (usually large mammals with whom we can empathise or identify). By contrast, in much environmental politics, or ecopolitics as Soper (1995) calls it, the subject of the future tends to be a collective one. The notion of future human generations is clearly used in anthropocentric environmentalist accounts; nature-possibly-devoid-of-the-presence-of-humans is the future's subject in some deep green ecopolitical positions. More recently, there have been a number of treatments which see the subject - indeed, the only real hope - of the environmental future as a heterogeneous collective composed of humans and nonhumans (see e.g. Cussins, 1997; Latour, 1998; Weldon, 1998).

Needless to say, the sort of 'subject' that is being deployed to carry the future does rhetorical work. Primarily, it invites us to identify with the subject. This is most obvious in the case of the individual sufferer, or with future generations of 'our' children. Further, such identifications - such identities - can function in a variety of ways. They can serve in the tacit reproduction of national identities, for example. As Samuel (1996) has documented, to analyse the future state of the Scottish environment in terms of the wildcat or the Caledonian forest is to situate oneself in relation to Scottishness. Or they can serve in the broad characterization of a public. In Callon's (1986) classic study of the Electricité de France's promotion of the development of an electric vehicle, the French public is represented as becoming environmentally concerned, but with a continuing investment in private transport.

One of the hoped-for effects is that such identification generates a sense of responsibility. We are responsible for this individual, or this collective, therefore we should do something to help them. However, as any rhetorical analysis will point out, to engage in such identification is to fail to identify with or differentiate oneself from another 'subject'. To be Scottish is not to be English. To be a public with a continued commitment to private transport (and hence, potentially, the electric car) is to be a public not interested in the improvement of pubic transport. To focus on the futures of individual sufferers is, possibly, to neglect implications for collectives. Central then to the analysis of the 'subject' of futures is then a sensitivity to whom or what gets marginalised. Needless to say, these marginalised identities can be transformed into counter-subjects of the future in the process of argumentation (as in Callon's (1986) case studies where Renault ripostes with a representation of the public's emergent interest in public transport).

Finally, and this is something I elaborate in the section on 'speed', the 'subject' of the future also connotes agency. In representations of the future, there is very often, if not always, a more or less developed narrative of how 'we', or some other constituency, should get to or avoid the future. Here, we would need to examine how the subject is represented as comporting itself in relation to the future. Portraying such a 'subject' as exercising an active agency, a purposeful striding toward the future, or caricaturing it as a passive actor being driven toward the future by circumstances beyond its control, will obviously have major implications.

Forms of Rationality: Substantive versus Instrumental

This complex dichotomy disguises a series of interlocking concerns only a few of which I will address here. Clearly, this dichotomy reflects a variety of aspects of the future that, in part, resonate with traditional sociological concerns. Thus one can view representations of the future in terms of the contrast between two classic (ideal typical) rationalities. On the one hand, there is the means-oriented instrumentalism critiqued by Weber and the Frankfurt School. Here, the ends entailed in the future are pretty much the same as those that inform the present. Thus, for example, foresight programmes (cf. Rappert, 1999; Webster, 1999) project present-day conditions - conditions which might be called fundamental or foundational - into the future. These might include the continuing need for economic competitiveness, the more or less undiminished power of the market (see Callon, 1998), the role of technoscientific innovation in gaining or sustaining economic advantage, the relation between nation states, global capital and globalisation and so on and so forth. Of course, alongside these assumptions are others concerned with the enduring nature of society (liberal democratic and neo-liberal capitalist, for example), of personhood (sovereign individual, for example – cf. Abercrombie, Hill and Turner,

1986), of nature (for example, resilient – cf. Schwarz and Thompson, 1990).

On the other hand, there are those representations of the future that are substantive, ends-oriented. These accounts can be more or less utopian explorations of possible futures in which the character of humanity and the form of society - that is, the meaning of the 'good life' - is examined. Obviously such explorations have a long lineage, not least in various fictional genres (see Manuel and Manuel, 1979; Kumar, 1987, 1991). But they also inform much social scientific thinking in one way or another. They can be couched in the terms of 'moderation' or 'realism' as Giddens (1990) does, or they can be formalistic utopias concerned with a possible future of untrammelled dialogue (e.g. instanced in Habermas' 'ideal speech situation'). They can be highly elaborated as in Marcuse's *Eros and Civilization* (1955) in which an ideal human pycho-historical condition (non-repressive de-sublimation) is posited as the realisation of happiness and freedom. Or they can be somewhat tacit or minimalist as in Marx and Engel's vision in the *German Ideology* (1964) in which multiple, disparate, freely chosen activity is the means to self-realisation. As they put it: 'a communist society there are no painters but at most people who engage in painting among other activities' (p.432). This suggests a remarkably contemporary de-centred self. It is contemporary because we find echoes of this fluidity in several accounts of postmodernity (e.g. Gergen, 1991). However, in contrast to Marx's view where fluidity is manifested in labour and practical activity, postmodern accounts (reflecting the 'linguistic turn') emphasise the fluidity of identity and of signification (cf. Michael, 1996 for an elaboration and exemplification of this analysis). In sum, we can see that utopias of various forms have played, and continue to play, a part in social scientific analysis. Most importantly, the part they play - by virtue of developing alternative models of the person, of the social, of nature that challenge existing versions - is that of critique. As Bauman (1976) notes, utopias (and specifically socialist utopias) serve several functions: they relativize the present by 'exposing the partiality of current reality, by scanning the field of the possible in which the real occupies merely a tiny plot, utopias pave the way for a critical attitude and a critical activity...' (p.13); they allow for the extrapolation of the present; they 'split the shared reality into a series of competing project assessments' (p.15); and they exert an actual influence on the course of history.

Now, this is not to say that instrumentally rational representations of the future do not incorporate critique. The critical focus of an instrumentally rational technique such as technological foresight is primarily upon means, processes by which to reach an end which remains largely 'uninterrogated', though, as Rappert (1999) demonstrates, such accounts can still entail an *implicit* recasting of substantive ends. For utopian critique the ends are of fundamental and *explicit* importance - the argument is openly over what it means to be a good person, to live the good life.

Having noted this, we might ask where does any particular utopia, and the critique it resources, come from? Is it not 'of its time' - contingent upon the historical circumstances in which it arose and reflecting the prejudices and predilections of its own epoch? This contingency of the utopian is what made Marx so reluctant to describe possible futures, and so scathing about the utopians. However, even so, with the proviso of contingency in place utopian thinking can nevertheless serve as a spur to uncovering some the assumptions that inform our common sense take on the world as demonstrated only too brilliantly in Donna Haraway's (e.g. 1997) work.

In light of the foregoing discussion, I need to ask how these different rationalities of the future work rhetorically. Perhaps most obviously there are the claims and counter-claims concerned with the realizeability of the future. Thus, advocates of 'instrumental futures' can claim they are being realistic, dealing with the way the world really is and will continue to be. They are thus making claims about the longevity of certain fundamental aspects of the present and, relatedly, the foreseeability of certain fundamental aspects of the future (e.g. the centrality of market). This 'realism' is a powerful rhetoric insofar as it can be turned on opponents who are dubbed 'dreamers' or 'optimists', and worst of all, utopians. But such rampant realism can also be a handicap insofar as it can reduce the sorts of options and possibilities available to, for example, a policy-maker. More generally, realism that attaches to instrumental rationality is itself up for grabs. Within these sorts of representations of the future, we can envisage a rhetorical game of 'more realistic than thou' where actors draw futures whose rhetorical potency (for some audiences at least) lies in their intense continuity with the verities of the present.

Of course, it is these very characteristics that render such instrumental futures so impoverished for 'utopians'. The charge that meets too much realism takes the form, in one way or another, of a poverty of imagination that encompasses both models of the present (people are selfish, the globalized market is unavoidable and so on) and of the future (ditto). Instead, the potential of people, or exemplifications of that potential, are called upon to warrant alternative, radically different views of what the future might be.

However, the story is never so simple. These rhetorics of the future crossover. Thus we have instrumental accounts which take on the trappings of utopian substantive futures and vice versa. For example, certain search engines are advertised with the promise of users' self-realization through the multiplicity and fluidity enabled by the internet. Conversely, environmental activists suggest that their 'utopian' futures are the only 'realistic' ones if we (the human race, the planet) are to survive. These crossovers of substantive and instrumental futures will clearly be shaped by whether the represented future is viewed as a good or a bad one. It is, then, to these dimensions of valency that I now turn.

Valency: Positive versus Negative

The future can be bad or good, positive or negative. It can be represented as a place of plenty and peace, or as a time of pain and privation. The sorts of futures that are attached to a given project are, as we might expect, often contested. Latour (1996) shows that the positive future attached to the projected transport system 'Aramis' - a positive future train system which can take the traveler directly to their destination without any stops en route - is countered by a negative future where the individual cars seating a few people are seen as exposing women to dangerous situations. These sorts of examples are innumerable. Representations of the future are from the outset engaged in a sort of pre-emptive argumentation over whether the projected state of affairs leads to good or bad. Such representations are rhetorically oriented toward stressing the good (by limiting discussion to those aspects of the projected future deemed 'important' or 'central') while downplaying the bad (by neglecting to discuss aspects deemed 'irrelevant' or 'tangential').

Good and bad futures can be pushed to the extreme. Catastrophic time is that of disaster - death, eruption, void. The imminence of catastrophe, according to Lingis (1998), which can range from the personal (e.g. a death) to the epochal (next ice age) through to the absolute (e.g. the explosion of the sun in 4 and half billion years' time - Lyotard, 1988), is contrasted to intelligible time. Intelligible time is modelled on work where we have expectations about how our efforts in the present bear fruition in the future. Representations of catastrophic futures can warrant a number of actions that range from pessimism and passivity to something altogether more liminal. As Lingis suggests catastrophic time has its attractions too - disorder, chaos, the possibility of new worlds - these lead to exhilaration, exultation. Desolation can yield intoxication.

However, let me return to less dramatic goods and bads. This rather obvious dimension of the futures neglects another set of rhetorics - what we might call 'attempted meta-rhetorics' - that can often come into play. Thus, representations of a good future can be charged with 'talking up' the future - that is, the enunciation of a particularly positive future can generate 'optimism' (or bullishness in markets). Similarly, a slight hint of negativity regarding the future can precipitate panic. These additional discourses do not, as it were, climb above to serve as commentaries upon representations of the future, they themselves comprise representations of the future.

Additionally, the good-ness or bad-ness of a future is not related only to the content of that future (such and such state of affairs) alone. It also relates to other aspects - for example, a future's foreseeability, or its predictability. A good future - whatever its substance - might be one which is seen to be of high certainty. Inevitably, as is only too well known, this dimension too is subject to controversy. Assessments (mathematisations, even) of uncertainty insofar as they are always contestable can be likened to the modality that Latour and Woolgar (1979) find in scientists' laboratory talk. Modality, as the level of certainty that is attached to the statement of a finding (or a future), affects the level of hostility with which such a statement is greeted (though see Lynch, 1985). Once scientific statements to the public were phrased in the vocabulary of certainty, once a particular technoscientific future could be presented with confidence. This was seen as 'reassuring'. Probabilistic accounts of the future can seem much less reassuring, can even suggest, to some at least, a dereliction of duty. And yet, in some quarters within the social sciences, modesty (as an admission of contingency and uncertainty) has become a major rhetorical tool. The point to this discussion is that the valency of a future is judged not simply in terms of its 'content', but also in terms of its 'form' (though putting it like this shows how these collapse into one another). Statements of (un)certainty can serve to sweeten a bitter prospect or make a sweet prospect bitter.

Speed: Slow versus Fast

If we envisage time as a line stretching ahead, if we imagine the future to be a point on that line more or less distant from us, at what we speed do we approach it? Writers such as McNaghten and Urry (1998) and Adam (1998) have reflected on the fact that there are emerging different times, different rates of change. In addition to clock time we now have an awareness of glacial time (incredibly slow environmental change) and nanotime (incredibly fast time associated the new information technologies). These speeds have implications for how we view the future, how we represent it. Indeed, the speed with which we approach a future can be a major part of the performativity of a representation of the future. For example, Rappert (1999) notes that a key part of technological foresight rhetoric is 'the assumption that change is taking place at an increasing pace' (p.532). Nowotny (cited in MacNaghten and Urry, 1998) sees the rise of 'instantaneous time' and the sense of increasing pace as affecting our views of the future to the extent we now see ourselves as living in an extended present. The future is always within our easy grasp and new rhetorics in which the future becomes immediately realiseable arise - MacNaghten and Urry (1998 p.150) quote the slogan 'I want the future now'.

However, this speed - this headlong rush, is not separate from the representation of particular futures. Different futures are associated with different speeds. Formulaically, we can put this point in the following way: if we move at Speed A to Future 1, on reaching Future 1 we will become enabled to travel at Speed B to Future 2, where B>A. In other words, getting speedily to a particular future will enable us to get even more speedily to a future future. There is here a process of increasing speed, acceleration, of progress. As many authors (e.g. Sachs, 1984, 1998; Virilio, 1977, 1995; Millar and Schwarz, 1998) have noted, this is the 'dream' of western modernity. Continual acceleration - constant improvement on past speeds - is a deeply embedded aspect of 'our' culture. No wonder, then, that it is so powerful a rhetoric. Indeed, this connection between speeds and particular futures finds an important expression in speculative representations of the far future (or what some would call science fiction). The greatest speeds can be achieved when we have no bodies: Virilio (1995) puts it thus: 'To expand, to dissolve, become weightless, burst, leave one's heavy body behind: our whole destiny could now be read in terms of escape, evasion' (p. 80; see also George, 1998, whose equation of speed, ubiquity and power readily evokes the ethos of disembodiment). In such fictions as the Star Trek series, the highest species - the most advanced, those who represent our potential evolutionary future - always seem to transcend corporeality, indeed materiality. They are pure energy and this is a higher state of being. The idea of disembodiment is thus attached to a notion of progress.

Here, we see that the potency of a rhetoric (in this case about speed towards a future) lies not 'merely' in its cultural rootedness - the ways in which it reflects well-grounded common-place arguments. Rather, it is embedded in a predilection with speed that is fundamentally tied to the development of certain technologies which have enabled the (gendered) corporeal 'pleasures' of high speed. This point has major implications and I will return to it in the concluding section of this chapter.

Finally, returning to issues raised in the section on the subject of the future, speed is also attached to subjecthood and agency. Speed is not, in this context, a matter only of quantity. There is a quality to speed too. As one moves toward the future, does one move willingly, or with passivity? Does one take up the controls of whatever vehicle speeds us to the future, or is one driven toward it? Do we resist passively, inertially? Or do we exercise willful resistance? These representations of the subject of speed obviously have much rhetorical mileage: low speed, or deceleration, can be negative (mindless inertia) or positive (reflexive obstruction). Similarly, high speed, or acceleration, can be negative (headlong, irresponsible rushing) or positive (entrepreneurial grasping of the future moment).

Concluding the Future: From Performativity to Prehension

The preceding sections have focused on representations of the future. In particular, I have attempted to unpick some of the parameters of these representations and to explore some of their rhetorical functions. To reiterate, I make no claim that the above list is exhaustive, or that the various parameters are mutually exclusive or even discrete. As we have seen at several points, there are crossovers between these different dimensions and any sustained rhetorical analysis of such representations of the future would need to consider these parameters in total, and much more systematically than I have been able to do here.

Having restated these provisos, I should point to another major limitation with the foregoing. My treatment of these representations of the of future has been just that - a consideration of *representations*: stories, characters, discourses, motifs, metaphors and so on and so forth. However, as mentioned in the introduction and in the section on speed, these representations are grounded in the material. The performativity of these representations does not take place in some abstracted, a-material domain. It is conducted in material settings, where bodies and texts, for example, come into contact or close proximity at least. As such these representations of the future connect actors who are at a distance in space and time. Is it possible to think about these performances, these connectivities, in a different way? Perhaps, in the way that for Mead, the 'present of the mind' is what connected the future and the past, performances are the present that connect, inter-relationally, the past and the future?

Let me get at these issues by posing the following questions: what is the 'present'? What are its borders, parameters, limits? To re-iterate, instead of thinking of the future as a representation, let us think of it as a 'textualization' by which I mean to connote that the representation takes material form. It is materialised on paper, on diskette, as binary signals that travel along wires and through airwaves. As such, we ask: when and where does a representation of the future 'seek' to have effects, that is perform? The 'textualised future' moves and in moving 'takes time'; it assumes a sort of extension in time. The 'present' of the performance is thus elongated. The parameters of the present can thus be said to be pliable, and, further, this pliability can be said to reflect the mode of movement, by which is meant (of course) technologies of communication. In sum, the representation of the future, conceptualised as a performative *materialised* artefact shapes the 'present' in which it performs.

In framing 'the future' in this way, I have, to some degree, modified some of the assumptions upon which the idea of 'lived temporal duration' is based. Specifically, in relation to the quote cited in the introduction ('the real past, just like the real future, is unobtainable for us, but through mind is open to us in the present'), I have rendered 'mind' relational and emergent. To the extent that a textualized future performs to us - that is, connects us to the 'origin' of that text - then the present is extended, encompassing the connection wrought by the 'performative textualised future'. Conversely, as 'producers' of textualized futures which we send out into the world, we are 'connected' to the receivers in an expanded present.

This model of the 'performative textualised future' derives from an emerging, albeit disjointed, perspective that includes such figures as Alfred North Whitehead, Michel Serres, Gilles Deleuze and Bruno Latour. It is fundamentally concerned with what Lash, Quick and Roberts (1998) in their introduction to the volume *Time and Value* describe as: 'a flattened and immanent scenario "of connections, fluxes, and objective intensities"' (p.4). Rather than time being stretched out linearly over space, and the present being a point on a line, or a 'moment' in the mind, it is emergent in the processes of connection and flux. Thus for Serres 'the present is singular and is seen as "a given assemblage of particularities"' (p.4). What makes the assemblage is a connection wrought by what Serres once called Hermes, and now calls angels. For Serres (1995), angels in their multitudes are a better metaphor for capturing the circulations and connections of multifarious, heterogeneous entities: humans, knowledges, languages, objects, and processes. As Serres puts it in response to the queston, 'why should we be interested in angels nowadays?':

> Because our universe is organised around message-bearing systems, and because, as message-bearers, they (angels) are more numerous, complex and sophisticated than Hermes, who was only one person, and a cheat and a thief to boot. Each angel is a bearer of one or more relationships; today they exist in myriad forms, and everyday we invent billions of new ones. However, we lack a philosophy for such relationships (1995 p.293).

Importantly, such connections entail contact and communication between disparate entities. For example, how is it possible that the same motifs appear in science and myth? How are these connected? As Latour (1987) and Harari and Bell (1982) note, Serres' approach to this 'parallelism' does not entail a dominance of one text or tradition over another - there is no critique or commentary or metalanguage exercised by one over the other (say, science over myth). Rather, there is seen to be what Latour calls a cross-over wherein the insights to be derived from myth or fables are no less valuable than those that flow from physics. One is not anachronistic, the other modern. They co-exist; they are co-present. For Serres there is no linear time, rather there are movements between order and disorder. Sometimes these are entropic, sometimes negentropic, sometimes cyclical, sometimes homeostatic. It is out of these movements that, we might say, time emerges. What connects, and renders order or disorder, is, as we have seen, 'angels'.

The 'performative textualised future' is just such an angel. It renders connections and in so doing, constitutes a present. But as noted above, such a textualised future by virtue of its materiality (and this, of course, applies to any textualisation) performs, and thus renders its connections, in part, through the medium in which it travels and the trajectory along which it journeys. The question we must now ask is: what form does this performance take? Or rather, how does this textualised future operate, work? And on what exactly? Here, we have to move away from thinking about performance as a process of signification, a process of having effects through signs alone, a process of hermeneutics (which, of course, bears the traces of Hermes' name). Rather, performance is heterogeneous - it operates on many levels, at once material and semiotic (and as Callon and Law, 1995, note, many such performances will be partly or wholly obscure to us). The implication is that the effectivity of these texts cannot be sought solely in their, albeit revocable, signification or meaning. We should not simply be seeking to excavate how such texts make connections through, for example, the making and taking of subject positions, or (even less likely) of the production of 'understanding'. Instead of 'comprehension', or even 'apprehension', we should, following Whitehead (1929) in his metaphysical philosophy of the organism, perhaps talk of 'prehension' (or what recently Latour, 1999, has referred to as 'articulation', and what Strathern, 1991, has called 'partial connection'). This term refers to the multitude of heterogeneous ways in which an entity (a category which includes humans) is 'attached' to, and emerges out of, the external world. Prehensions have a vectorial character and involve 'emotion, and purpose, and valuation, and causation' (p.25). Indeed, an actual entity is simply what Whitehead calls a 'concrescence' (or coalescence into some-thing concrete) of prehensions. For each prehension, there are 'three factors: (a) the 'subject' which is prehending, namely the actual entity in which that prehension is a concrete element; (b) the 'datum' which is prehended; (c) the 'subjective form' which is how the subject prehends that datum' (p.31). The performative textualised future can be regarded as deeply embroiled in such prehension - it is a 'datum' that works on a variety of levels - purposive, emotional, and even causal. As such, it is not prehended by a 'subject' or enters 'subjectivity' as traditionally understood (and Whitehead does signal this expanded view of 'subject' with his liberal use of scare quotes). Of course, what is also being tacitly drawn into this picture of prehensive performativity is not only the materiality of the message, but that of the senders and receivers: the corporeality of humans (albeit a contingent one) is thus incorporated into this account. On this score, at least, Whitehead's metaphysics serves as a precursor of the (a)modern admixtures of human and nonhuman that we find theorised through such concepts as hybrid (Latour, 1993), monsters (Law, 1991), and cyborgs (Haraway, 1991).

In sum, then, prehension has the double meaning, ambiguity even, that an angel (or an association in actor-network theory) does. It connotes both a 'message' that travels to an existent 'subject' (or receiver or actor) and a 'message' that partly (re)constitutes the 'subject', that serves in its becoming. The point of course is that performative textualised futures as messages, that is to say 'angels' and 'data', contribute to this process.

How to study this impact - the performance - of these textualized futures is still very much up for grabs. It requires, I suspect, a sensitivity to partiality and prehension that we're only beginning to develop. However, to make matters even more complicated, surely we would also want to reflect on how such performances entail both, or rather, indissolubly, the physical movements of texts, and the signifying content of the represented futures. Performance here, of course, straddles the material and the semiotic and, in analysing it we would want to consider cogently both the representations of the future and textualized performative futures, that is both content and form. In a way, we would be presiding over the collapse of this dichotomy. As we chart how represented futures finesse, interrupt, as well as sustain, the relations mediated by material flows of texts, as we explore how a particular version of the future is affected in its performance by the medium through, and the trajectory along which it is delivered, we begin to unravel the interwoveness of the material and the semiotic, and the past, present and future. By way of fleshing out how this unravelling might proceed let me pose a few questions. Taking the example of xenotransplantation (XTP), we can ask: What are the representations of the future circulated by advocates of XTP? What are the trajectories along which these travel (e.g. press releases, policy documents, scientific papers)? How do these cross-cut other representations of the future concerning, for example, the need to improve the public understanding of biotechnology (presented in the form of Eurobarometer results)? How do these two performances of XTP interact? How are these further complicated by the performance of XTP futures in fictional genres like science fiction films, for instance (Brown, 1999)? From this simple example, we can readily list a range of genres (e.g. documentary news, social scientific, science fictional), an array of media (e.g. press releases, films, government reports), and a selection of trajectories (between scientists and publics, scientists and legislators, fiction writers and consumers). Moreover, weaving in and out of these circulations of textualized performative XTP futures will be further complexifying narratives about, for example, the future of democracy, citizenship and consumption. To the extent that these different media partially connect – form prehensions - with others, to the extent that these trajectories meet, they 'ravel' at a common locus that constitutes the 'present'. In this 'present', XTP (and other) futures intertwine – 'concresce' in Whitehead's terms - in a complex performance that it is our task to 'un-ravel'. All I have done here is haltingly attempt to formulate this task of unraveling textualized performative futures. Inevitably, actually tackling such tasks lies in the future.

References

Abercrombie, N., Hill, S. and Turner, B. (1986) *The Sovereign Individual*, Allen and Unwin, London.

Adam, B. (1990) *Time and Social Theory*, Polity Press, Cambridge.

Adam, B. (1998) *Timescapes of Modernity*, Routledge, London.

Bauman, Z. (1976) *Socialism: The Active Utopia*, Allen and Unwin, London.

Billig, M. (1987) *Arguing and Thinking: A Rhetorical Approach to Social Psychology*, Cambridge University Press, Cambridge.

Billig, M. (1991) *Ideology and Opinions*, Sage, London.

Billig, M., Condor, S., Edwards, D., Gane, M., Middleton, D. and Radley, A. (1988) *Ideological Dilemmas*, Sage, London.

Brown, N. (1998) *Ordering Hope: Representations of Xenotransplantation - an Actor/Actant Network Theory Account*, PhD Thesis, Centre for Science Studies, Lancaster University.

Brown, N. (1999) Xenotransplantation: Normalizing Disgust, *Science as Culture*, 8, 3, 327-355.

Callon, M. (1986) The sociology of an actor-network: the case of the electric vehicle, in M. Callon, Law. J. and Rip, A. (eds), *Mapping the Dynamics of Science and Technology*, MacMillan, London, 19-34.

Callon, M. (ed) (1998) *The Laws of the Markets*, Blackwell, Oxford.

Callon, M. and Law, J. (1995) Agency and the hybrid collectif, *The South Atlantic Quarterly*, 94, 481-507.

Cussins, C. (1997) Elephants, biodiversity and complexity: Amboseli National Park, Kenya, paper presented at *Actor-Network Theory and After Conference*, Keele University.

Freeman, M. (1993) *Rewriting the Self: History, Memory, Narrative*, Routledge, London.

George, S. (1998) Fast castes, in Millar, J. and Schwarz, M. (eds). *Speed – Visions of an Accelerated Age*, The Photographers' Gallery and the Trustees of the Whitechapel Art Gallery, London, 115-18.

Giddens, A. (1990) *Consequences of Modernity*, Polity, Cambridge.

Giddens, A. (1991) *Modernity and Self-Identity*, Polity, Cambridge.

Harari, J.V. and Bell, D.F. (1982) Introduction, in Serres, M. *Hermes: Literature, Science, Philosophy*, Johns Hopkins University Press, Baltimore, ix-xl.

Haraway, D. (1991) *Simians, Cyborgs and Nature*, Free Association Books, London.

Haraway, D. (1997) *Modest_Witness@Second_Millenium.FemaleMan.Meets OncoMouse: Feminism and Technoscience*, Routledge, London.

Kumar, K. (1987) *Utopia and Anti-Utopia in Modern Times*, Blackwell, Oxford.

Kumar, K. (1991) *Utopianism*, Open University Press, Buckingham.

Lash, S., Quick, A. and Roberts, R. (1998) Introduction: Millenniums and Catastrophic Times, in Lash, S., Quick A. and Roberts, R. (eds), *Time and Value*, Blackwell, Oxford, 1-15.

Latour, B. (1987) Enlightenment without Critique: A Word on Michel Serres' Philosophy, in Phillips, A. (ed), *Contemporary French Philosophy*, Cambridge University Press, Cambridge, 83-97.

Latour, B. (1993) *We have Never been Modern*, Harvester Wheatsheaf, Hemel Hempstead.

Latour, B. (1996) *Aramis, or the Love of Technology*, Harvard University Press, Cambridge, Mass.

Latour, B. (1998) Modernise or Ecologise? That is the Question, in Braun, B. and Castree N. (eds), *Remaking Reality: Nature at the Millennium*, Routledge, London, 147-187.

Latour, B. (1999) *Pandora's Hope: Essays on the Reality of Science Studies*, Harvard University Press, Cambridge, MA.

Latour, B. and Woolgar, S. (1979) *Laboratory Life: The Social Construction of Scientific Facts*, Sage, London.

Law, J. (1991) Introduction: Monsters, Machines and Sociotechnical Relations, in Law, J.(ed.), *A Sociology of Monsters*, Routledge, London, 1-23.

Law, J. (1994) *Organizing Modernity*, Blackwell, Oxford.

Lingis, A. (1998) Catastrophic Times, in Lash, S., Quick, A. and Roberts, R. (eds), *Time and Value*, Blackwell, Oxford, 16 - 31.

Lynch, M. (1985) *Art and Artifact in Laboratory Science*, Routledge, London.

Lyotard, J.F. (1988) *The Inhuman*, Polity, Cambridge.

MacNaghten, P. and Urry, J. (1998) *Contested Nature*, Sage, London.

Manuel, F.E. and Manuel, F.P. (1979) *Utopian Thought in the Western World*, The Belknap Press of Harvard University Press, Cambridge, MA.

Marcuse, H. (1955) *Eros and Civilization*, Vintage Books, New York.

Marx, K. and Engels, F. (1964) *The German Ideology*, Progress Publishers, Moscow.

Michael, M. (1996) Pick a Utopia, Any Utopia: How to Be Critical in Critical Social Psychology, in Parker, I. and Spears, R. (eds.), *Psychology and Society: Radical Theory and Practice*, Pluto Press, London, 141-52.

Millar, J. and Schwarz, M. (1998) Introduction - speed is a vehicle, in Millar, J. and Schwarz, M. (eds), *Speed - Visions of an Accelerated Age*, The Photographers' Gallery and the Trustees of the Whitechapel Art Gallery, London, 16-21.

Plein, L.C. (1991) Popularising Biotechnology, *Science, Technology and Human Values*, 16, 474-90.

Potter, J. and Wetherell, M. (1987) *Discourse and Social Psychology: Beyond Attitudes and Behaviour*, Sage, London.

Rappert, B. (1999) Rationalising the Future? Foresight in Science and Technology Policy Co-ordination, *Futures*, 31, 527- 45.

Sachs, W. (1984) *For Love of the Automobile*, University of California Press, Berkeley.

Sachs, W. (1998) Speed limits, in Millar, J. and Schwarz, M. (eds), *Speed - Visions of an Accelerated Age*, The Photographers' Gallery and the Trustees of the Whitechapel Art Gallery, London, 123-132.

Samuel, A. (1996) *Science as Practice: Conserving Scotland's Natural Heritage*, PhD Thesis, School of Independent Studies, Lancaster University.

Schwarz, M. and Thompson, M. (1990) *Divided We Stand,* Harvester Wheatsheaf, Hemel Hempstead.

Serres, M. (1995) *Angels: A Modern Myth*, Flammarion, Paris.

Soper, K. (1995) *What is Nature?,* Blackwell, Oxford.

Strathern, M. (1991) *Partial Connections*, Rowman and Littlefield, Savage, Maryland.

Virilio, P. (1977/1986) *Speed and Politics*, Semiotext(e), New York.

Virilio, P. (1995) *The Art of the Motor*, University of Minnesota Press, Minneapolis.

Webster, A. (1999) Technologies in Transition, Policies in Transition: Foresight in the Risk Society, *Technovation*, 19, 413 - 21.

Weldon, S. (1998) *Runway Rhetorics and Networking with Nature: A Study of Scientific Expertise in Environmental Impact Assessment*, PhD Thesis, School of Independent Studies, Lancaster University.

Whitehead, A.N. (1929) *Process and Reality,* Cambridge University Press, Cambridge.

Part Two

Language and the Social Rhetoric of Technical Futures

Part Two

Language and the Social Structure
of Everyday Practice

3 Forceful Futures: From Promise to Requirement

HARRO VAN LENTE

While we all know that technical change is part and parcel of ongoing social and economic changes, technology often appears as an autonomous force. Indeed, one of the striking things about technological futures is that they often appear in the imperative mode. That is, once defined as promise, action is required. Statements about future technological performance are not received as factual descriptions to be verified or falsified in due course. Instead, they mobilise attention, guide efforts and legitimate actions. Technological futures, are then forceful and render technological developments a specific dynamic (Van Lente, 1993).

While each particular development builds up its own credibility (Latour, 1987; Callon, 1999), forcefulness is also derived from a common background. Arguments and urgencies that shape specific projects draw upon widely shared metadiscourses about social change and technological developments. Such metadiscourses have their own history and dynamic and provide the terms and arguments that are indispensable for any innovation. In this chapter[1] I will explore the force and dynamic of this common background. In particular, I will investigate how technological futures are embedded in well-established vocabularies. A key element here is the notion of normative technological progress that cannot or should not be stopped. I will show how this notion and the concomitant social arrangements fuel the force of technological futures and how this shapes the dynamic of technological developments. The argument will be illustrated through a potted history of High Definition Television in the early 1990s.

Philosophers and historians have been interested in 'technical progress' as a central element of modern western society, both as a historic force in the shaping of modernity, and as a widely shared cultural value. With all their differences, 18th-century philosophers like Condorcet, Comte and Smith agreed on the promise of technical progress for the improvement of society. Historians have described the history of industrialising Europe in the 19th century as an implementation of the promise of technology, or in Landes' (1969) terms, of *The Unbound Prometheus*. Most scholars writing about 20th-century civilisation will agree that Rationality, Progress and Technology are as important as the other 'big words' of modernity: Freedom, Equality and Brotherhood.[2]

The prominent place of technological progress in western culture has been studied in varying ways but widespread is the tendency to analyse, or, if you like, unmask, the powerful place of technical progress as a 'belief' in progress (Van der Pot, 1985). Other authors have pointed to seemingly autonomous technology, and attempted to show that it is really the outcome of interest politics (Winner, 1986). Be that as it may, what should be looked into is how this so-called 'belief' in technical progress is maintained, and what it entails for concrete dynamics of development.

Another way to grasp the apparent power of 'technical progress' is to study it as a language phenomenon with concomitant social consequences. Language, of course, is by definition part of the common background in which particular instances of technical change are embedded. Moreover, the advantage of studying language use over studying convictions or particular beliefs is that the former is accessible to the analyst, while beliefs and convictions are notoriously difficult to examine. The question, then, is what kinds of language strategies are involved and how these affect the fate of technical futures. The direction I propose to take is to address 'technological progress' as an *ideograph*, that is, as a well-embedded but relative empty term that due to its flexibility helps to mobilise support and suggest, or even enforce, specific directions.

Ideographs

Two decades ago the linguist McGee introduced the term ideograph in an effort to escape from of the deadlock in which discussions on the relation between ideology, power and social control tended to arrive (McGee, 1980). It is an alternative way of understanding what collective conviction means. Does a collective have, or not have, its own 'mind' or 'belief', from which the individual can, or cannot, escape? McGee tries to hold to a middle of position, by stating that people in a collective behave according to this collective, but not to the extent that they are determined by ideology. The point is that we do not necessarily follow the ideology. We are not socialised in a way a dog is trained, or in the way a child is forced to speak its 'mother tongue'. Instead, we are handed a vocabulary.[3]

Human beings are 'conditioned', not directly to belief and behavior, but to a vocabulary of concepts that function as guides, warrants, reasons, or excuses of behavior and belief (McGee, 1980 p.6).

Concepts such as 'freedom', 'property', 'privacy' or 'democracy' function as ideographs. McGee offers the following tentative definition:

> An ideograph is an ordinary language term ... a high order abstraction, representing collective commitment to a particular but equivocal and ill-defined normative goal. It warrants the use of power, excuses behavior and belief ... and guides behavior and belief into channels easily recognized by a community as acceptable and laudable (McGee, 1980 p.15).

An advantage of an analysis with ideographs is that the analyst does not have to assume a value or belief to be at work. Hence, actors themselves use ideographs and the analyst can trace their effect. Characteristic of these concepts is that they imply a *commitment*. The phrases in which they are used, can, when looked at superficially, be read as logical statements. In fact, however, they are not so much arguments with a logical structure, as invitations for political action. 'Ideographs signify and 'contain' a unique ideological commitment' (McGee, 1980 p.7). Proposals for public action and public legitimation are in the long run based on such words. Ideographs are the baseline of public and political discussions, and they are self-sufficient.

The reason that ideographs can fulfil such a special role, is that their specific contents can vary. For a liberal, 'freedom' means something different than for a social democrat or a Marxist. One can find such a confusion of speech upsetting, and try to introduce some order by giving one's own definition, or - somewhat less ambitiously - by neatly cataloguing the differences. In both cases one overlooks the fact that this *flexibility*, this relative openness, in fact harbours the force of these terms.[4] In fact, the flexibility of an ideograph enables *linkages* to be made with earlier instances of the ideograph or with other ideographs. This is most clear when one investigates how ideographs are used in mobilising support. I will discuss 'vertical' linkages which fill up the ideograph with instances from the past, first, and show how often mobilisation work for specific technologies partakes of this strategy.

Strategies of Ideographs

Generally, if someone wants to defend an issue, he or she can underline its importance by referring to comparable cases out of the past which teach generally accepted lessons. If a Dutchman wants to defend the rebels in a Central American country, he can refer back to the Dutch struggle for independence against the Spanish yoke. The struggle for 'independence' now is brought into line with the struggle for 'independence' then. The audience knows the example and understands the lesson. That is, the rebels are fighting for a good cause and they deserve support. Someone else might want to emphasise the need to respect 'the law' and, for that purpose, compare the rebels with, say, members of a terrorist group. The audience will be induced to condemn the rebels as unlawful and to dismiss their actions as counterproductive terrorism. Breaking 'the law' now is brought into line with breaking 'the law' then. In both cases the plea ends at a point which cannot be defended any further, the ideograph 'independence' and the ideograph 'respect for the law', respectively. In both cases parallels are drawn with exemplars, i.e. well known historic events with an accepted lesson.

My point here is not that historical evidence is used selectively; the point is that a forceful plea must be embedded in history, in a narrative of which the lines and lessons are recognisable and interpretable for the audience. From the bridge that the ideograph builds between past and present, *implications for action* follow, whether these are supporting or fighting the rebels. Different events that are supposed to be exemplars for the present case are linked by means of the ideograph. And thanks to the ideograph clear lessons can be distilled from history.

Next to ideographs such as 'independence', 'the law', or 'freedom', there is an ideograph of *'technological progress'*. 'Technological progress' is also an abstraction that expresses a social commitment, and an appeal to this term can legitimate proposals and activities. It is not the content of this concept which makes this possible - for the content is flexible. Rather, it is due to the term's capacity to connect diverse historical situations and, by doing so, to draw implications for action. Similarly, Brown's chapter in this volume shows how the interpretatively flexible temporal abstractions of breakthrough and discovery operate by evoking previous historical objects whilst also mandating future ones.

Our cultural repertoire indeed offers many bits of history that are exemplary of 'technological progress'. All of us are familiar with the miraculous multiplication of power by means of the steam engine, the boons of Pasteur's inventions, or the importance of the voyage to the moon. Once support is sought for a concrete technological development, this repertoire can be used. Stephen Hill, in his *Tragedy of Technology* (1988), gives various examples of reading history as a proof of technological progress. Another example is the way in which proponents of the Human Genome Project compare the project with the adventure of putting a man on the moon, thereby giving it the same urgency.

In the history of High Definition Television (HDTV), and we will come back to this story later, one can see the same rhetorical strategy. A report of the United States Office of Technology Assessment (the OTA is a former advisory agency of the American Congress), illustrates the importance of HDTV by drawing parallels with important steps in history.

> As early as 1883 inventors dreamed of transmitting visual images. By the 1920s, significant efforts were under way to scan and project images. ...TV was still futuristic at the time of the New York World's fair in 1939 but finally erupted into widespread commercial use in the 1950s. ... Television technology is now on the threshold of a new evolution. We are on the verge of combining digital based computer technology with television... The impacts of the development of HDTV will ripple through the US economy (OTA, 1990 p.iii).

Likewise, the British magazine *New Scientist* (1992) states:

> In the early summer of 1889, George Eastman, founder of the Kodak company, provided Thomas Edison with the first batch of film for his experiments with a cinema and projector. One hundred years later the electronics industry is gunning for film. It wants to replace film with magnetic tape that it can reproduce more cheaply and more conveniently the same high-definition pictures, or images of great clarity and detail, on giant TV and cinema screens. The commercial prospects for the new medium are enormous.

In these citations HDTV is put into one category with the inventions of television, photography and film, respectively. The same happens in another characterisation of HDTV, namely of HDTV being the 'next generation television', the next step after black and white television and colour television. The OTA report says, for instance:

> As an entertainment medium, HDTV is not revolutionary. It is simply another step in the ongoing evolution of television that began with black and white (B&W) TV in the 1940s and will continue into the future with as yet undreamed of technologies (OTA p.49).

The point of these quotations is the ease with which authors can take up historical instances as predecessors, and the message that they hope their strategy will carry. They can use the past because 'technical progress' is seen to be of one piece. What happened as early as 1883 is as much technical progress as that which we are undertaking now. The ideograph 'technical progress' thus contains an element of ongoing evolution, an unfolding logic that is captured in the notion of 'a next generation'. The notion is part of the rhetoric of a progress that should not be stopped - because not to have the next generation is to commit collective suicide. While the inevitability of technical progress as such is now contested, for specific technologies this cultural mould offered by the ideograph remains available, and is used without any need for justification.

The Weight of Ideographs

The force of the ideograph 'technical progress' is not independent of the status of other ideographs. A culture will have a cluster of ideographs, which are defined in relation to each other. Within this cluster, ideographs have unequal weights. The hierarchy of weights can change, but this requires effort. In this respect we could speak of a synchronic or horizontal dimension of ideographs, to be distinguished from the diachronic or vertical dimension that organises different pieces of history. It is in the horizontal dimension that the effort is exerted to alter the relative weight of one ideograph in relation to other ideographs.

> [W]hen we engage in ideological argument, when we cause ideographs to do work in explaining, justifying, or guiding policy in specific situations, the relationships of the ideographs changes (McGee, 1980 p.13).

Changing the relations within a cluster of ideographs is a pre-eminently political action. McGee gives the following example. When Nixon during the Watergate affair was pressed hard by accusations of violating the law and refused to deliver certain documents, he did not attack the ideograph 'the law'. Instead he tried to increase the relative weight of other ideographs:

> Before a mass television audience Nixon argued that a President's conversations with advisers were entitled to the same privilege constitutionally accorded exchanges between priest and penitent, husband and wife, lawyer and client. ...he magnified 'confidentiality' by praising the ideograph as if it were a person, attempting to alter its 'standing' among other ideographs, even as an individual's 'standing' in the community changes through praise and blame (McGee, 1980 p.13).

For the ideograph 'technical progress' the same possibility holds. It can be praised in order to increase its standing in relation to other ideographs. In that case a plea is held for technological progress *in general*, independent from concrete developments. Support is not sought for a specific technology, or a specific sector, but for 'technological progress' or 'technology', often in relation to attempts to alter the relative importance of other ideographs.

In the 19th century and the better part of the 20th century, 'technical progress' was praised as such. Especially in rapidly modernising societies like the USA and the Soviet Union (in the first decade), the ideograph became a cornerstone of the culture, from the motto of the Chicago World Fair ('Science Finds, Industry Applies, Man Conforms') to the way Vannevar Bush (1945) referred to 'Yankee ingenuity' as the mainstay of America, which had only to be supported with more science to actually create the 'endless frontier'.

By now, more arguments have to be given when one praises 'technical progress'. But these arguments are typically part of a struggle with other ideographs. One very common tactic is to define a counter-ideograph and condemn this counter-ideograph and those who follow it. It is interesting to observe that the force of a condemnation is generated through the tension between one's own ideograph and its antonym: emotional aversion and irrationality, refusal to enjoy present and future benefits of technology - these are all bad but only in terms of the ideograph of 'technical progress'. If you are defending another ideograph, they lose their force. The defenders of the other faith, e.g. the principled critics of technology, will take the same approach in condemning the ideograph of 'technical progress', and appear as inconclusive to the technological optimists as these to them.

Other actors make use of other ideographs, such as 'democracy', 'solidarity', or the converse of the technological progress ideograph: the 'dangerous technology'. Technology as a bane, rather than a boon. The converse ideograph is not something of recent years. In fact, besides the tradition where technology is praised as the benefactor of humanity, our culture also knows a tradition of visions in which technology leads to insanity, destruction and downfall. Opposite Bacon, Edison, and Pasteur stand Frankenstein's monster, the atom bomb and Three Mile Island.[5] Opposite 'technological progress' stands 'dangerous technology' as anti-ideograph. The former is available for those who want to stimulate a concrete technological development, the latter for those who want to protest against it.

Specific technologies become part of the struggle between ideographs, because they offer themselves as a temporary focal point (as with nuclear power, and more recently, with genetic engineering), and because they refer to promises and threats that are part of a general struggle. New medical technology will conquer even more illnesses, biotechnology will supply humanity with more food, and nuclear fusion will finally provide us with abundant energy. Following this line of reasoning, stopping or postponing the concrete development is undesirable, because it will hamper progress in general. Opponents of concrete developments are, in this light, morally objectionable. They are cynical cultural pessimists. How could one possibly not want medicine to progress and technological potentialities to be realised?

How can one want not to solve the world food problem or energy shortage? In the recombinant-DNA debate a recurring argument was that each day that recombinant-DNA research was delayed people would die unnecessarily. Or, in an ironical formulation:

> How much do we need Recombinant DNA? Fine, we can do without it. We have lived with famine, virus and cancer, and we can continue to.[6]

If one accepts the message, recombinant DNA technology must proceed, because the progress it brings is more important than some other issues. Again, there is an element of self-justification, now because of how acceptance of a specific technology is joined to acceptance of technology in general, with mutual reinforcing.

Mandate to Technologists

The next step in the exploration of the force of technological futures is to note that a 'good technologist' has certain *tasks*, just like the 'good citizen' in a democracy has certain duties, e.g. voting. The tasks of the technologists are not only to maintain existing technological systems. The technologists must first and foremost have an open eye for what is new, and be alert to questions like: Are there technological possibilities that can be employed to make existing systems more efficient? Which developments are going on, or can be expected, and which of these are promising? What could be the successor of the present system? In which time-span is this succession to be expected, or, in other words, when is the present technology rendered out of date? On which specifications of this new system should I orient my work? Will the new technology arrive in time?

How this grouping of '(good) technologists' with well-defined tasks could emerge is a question of social dynamics, a question of the division of labour at the societal level. Clearly, discussing the question in any detail is beyond the scope of this chapter. I will limit myself to the claim that social dynamics relate to the rise of the ideograph of 'technical progress',[7] and that the resulting division of labour between 'technologist' and the rest of society has the character of a *mandate*. It is not possible, within the scope of this chapter, to argue this rather sweeping view in necessary detail. A few tips of the iceberg of observations on and analysis of historical complexity must suffice to make the points plausible.

A starting point is the functional divide in our society between *maintaining* and *inventing* technology, which relates, for instance, to John Staudenmaier's (1985) distinction between the 'maintenance' and the 'design' constituency. The maintenance of the present system belongs mostly to the task of the technical, or blue-collar workers. Orientation on and design of the new system is a different task; a task for (those whom I have called) 'technologists'. This division of labour was already noted by Ortega Y. Gasset in his essay *Man the Technician* (1962), an analysis which has unfortunately been neglected. Following him, the history of technology can be divided into three stages: primitive, traditional and modern. In the primitive situation, technology is not recognised as such. As anthropological studies show, it is one with the natural state of things. Technology is not something independent, let alone that a systematic improvement of technology could be pursued. In traditional society, as a result of social differentiation, technology is seen as an independent entity: making shoes is a task of the shoemaker, the baker bakes bread and the architect builds temples or cathedrals. The trade is passed on from generation to generation, from master to apprentice. In this situation there is no space for systematic inventing. Improvements are made to procedures and processes, but these are sooner seen as indications of someone's skill and personal style, than that they are presented as new discoveries. The technical is not separated from the person who practices it, the artisan. The practice of the technology, and the process, are hidden behind the person. The task of the artisan is to fulfil functional needs - a task that continues in the self-image of the modern technologist. In the modern period a further differentiation takes place. Inventor and worker are separated, where they were once united in the artisan.[8] The machine - the tool that works by itself - is an exemplary illustration being handled by a worker but invented by a technologist. This functional divide goes together with the idea of technology *as such*.

> Today technology stands before our mind's eye... And this enables certain persons, called engineers, to devote their lives to it. ...he knows before he begins to invent that he is capable of doing so... The engineer need not wait for chances and favourable odds; he is sure to make discoveries. How can he be (Ortega Y. Gasset, 1962 p.144)?

The technologist is certain that s/he will make discoveries. S/he lives in the (modern) certainty that inventions will be done, that is, in the certainty that there will be something like technological progress. This is not just a personal · conviction. Technologists can live this way because society expects this from them, and lends them the space to work at inventions.

Note that due to this division of labour between worker and inventor/technologist that Ortega stipulates, technological promises have become an inherent part of technology. The differentiation not only makes technological promises possible - promises presuppose the possibility of purposeful invention - but also necessary. In contrast, the practice of artisans is based on existing, visible products, and so is their credibility. A customer who orders something knows which products will be delivered, practically the same products that have been delivered before. On the contrary, the product of the technologist, the invention, is localised in the future. At the moment of 'ordering' it cannot be seen or demonstrated, it can only be promised. Technologists not only live in the certainty of invention, but also in the promise of invention. The *raison d'être* of technologists is the technological promise.

The broad sweep of historical diagnosis, traced with the help of Ortega, can be filled in, and modified, for particular situations and developments. Illustrative of this is a case study of how the still uncertain grouping of what came to be the electrical engineers related themselves to the new electrical technology at the end of the last century (Marvin, 1988). This example, discussed below at greater length, shows how the engineers sought to establish an identity and a social space by positioning themselves not just as producers of services, but also, and particularly, as experts of promises. In general, exclusionary strategies can be recognised in all professions and attempts at professionalisation, because of the interest in a professional monopoly. The interest of this example for my argument is that it highlights the dependencies of technologists on the audiences that have to be receptive to their promises.

The 'electrical engineers' in the United States toward the end of the last century were searching for a stable identity (Marvin, 1988). Their strategy was to exclude others and by presenting the public as naive - a strategy that is still successfully practised by technologists today. In the United States of the last century, the young and undefined group of electro-technologists impressed upon their audience that progress is served by electricity. Even more, electricity is essential for progress. This became a widespread conviction in the early decades of the 20th century in the USA. The general promise of electricity justified the social existence of an electrotechnical profession. Magazines such as *Electrical Review* and *Electrical World* thus carried a message to the broad public. That is, that everything can be better and more beautiful with electricity and with electrical engineers, of course.[9]

High expectations were raised for electricity, sometimes too high to be realized by the electrotechnical experts. A strategy was needed to get out from under these claims without losing the faith of the public. The solution was to have the differentiation between experts and outsiders (which is always necessary for the development of a profession) also apply to promises. It was considered technologically *naive* to harbour expectations which experts saw as unrealistic. The professional magazines could make fun of the lay public that expected too much of future and existing electrical technology.

In respect to the early days of telephony, for example, the following anecdotes were recounted in their pages: the labourer that crams a piece of paper into the horn and waits for a written answer; the farmer who, without dialling a number calls into the horn that his sister must quickly come home because his mother is ill; the office boy who keeps on nodding politely when he is asked whether he is there, and does not catch on that the person on the other side of the line cannot be satisfied with that. By picturing others as naive, electrotechnicians could safeguard their monopoly of expectations about present and future technology.

Thus experts had to tone down expectations, but not too much, because the expectation that electrification would bring wonders was exactly the basis of their existence. So the experts gave themselves the task to teach the public 'proper' promises. Carolyn Marvin, who studied this part of the history of electricity, introduced the notion of promises as currency:

> One official boundary at which electrical insiders and outsiders met was negotiated in a currency of promises given by insiders to outsiders ... Expert and popular literature alike monitored the rhetoric of reciprocity, watchful for any breach in the vague but binding bargain between experts and their publics in behalf of technical progress (Marvin, 1988 p.16).

Eventually, the electrical engineers sought and found a social domain, or space for themselves. They got, in other words, a *mandate* to deliver electrical progress; they sought and found a mandated territory. An important part of this was their linkage with the strong rhetoric of the ideograph of 'technical progress'.

In a more diffuse way, technologists have obtained a similar mandate. The 'vague but binding bargain between experts and their public' is a general characteristic of the social locus of technologists. In a sense, technologists are defined by this implicit social contract, and their identity and self-image are related to it. Developers of new technology can demand a space for their activities in the name of 'technical progress', and because they are doing so, because they are serving 'technical progress', they are (good) technologists. Generic and specific technological promises are the currency in which the transactions in this social contract are concluded. Technologists have a space, a mandated territory, within which they count as experts. They are the ones who are allowed to speak first, they can in the first instance determine what is to happen. That they are allowed to have this space can be legitimated by a claim referring to the ideograph of technological progress.

In my argument, mandate, or mandated territory - a social contract specifying legitimated space for more or less autonomous activities and obligation to fulfil the tasks involved well - has been linked to the rise of one or more ideographs. One can see similar long-term developments for other ideographs than 'technical progress'. For example, for the ideograph 'health', where the medical profession has quite some freedom to act according to their own judgement. Or 'justice', where the judges have the right to speak. These territories are historical achievements and their boundaries are not set beforehand. In these examples, the medical profession, and the different professions involved in the law, have probably been the main determinants in setting the boundaries. In the case of technology, there is no such clearly recognisable social group, and the identity of the 'technologist' has emerged together with the rise of the generic promise of technology.

The boundaries and the nature of a mandate are always debatable, and this implies a struggle of ideographs. Disputes on mandates can be found for all the professions, e.g. alternative treatments questioning the monopoly of the medical profession, and the limitations of established medicine. In the case of technology, boundaries are also debatable. This can be seen very clearly in the contested nature of technology as such.

It is exactly at times when boundaries are threatened, that *spokespersons* come to the fore to speak in the name of the mandated territory. They are enthusiastic, concerned, encouraging or reproachful in name of 'health', 'justice', or 'technological progress'. Captains of industry or actors within government thus present themselves as spokesperson for 'technological progress'.

The mandate to technologists has a dual nature. As a complement to freedom, there is the obligation to take care of the territory. If the task to guard over 'technological progress' is neglected, they can be confronted with this and called to account. This is especially clear when a technological promise is in danger of not being pursued sufficiently. The technologists themselves, then, are also blamed, and this can only happen because all the parties share the idea that technologists have the obligation to make technological progress happen.

Self-justifying Technologies

What are the implications of 'ideograph' and 'mandate' for actual technological developments? The history of High Definition Television in the early 1990s is a good site to explore these issues because the promises and expectations have had a readily accessible high public profile. But since the argument is more general, I have drawn evidence from other cases as well.

At this point, a brief history of HDTV is helpful.[10] Before 1985, HDTV was held to be a promising idea within the domain of technologists but it was not widely promoted, say as a national cause, except, perhaps, in Japan, where NHK, the Japanese national public broadcaster, succeeded in co-ordinating a lot of research on HDTV from 1968 onwards. The CCIR meeting in Dubrovnik in May 1986 marks the transition. At the meetings of the International Radio Consultative Committee (CCIR), that are held every four years, global standards for television production and broadcasting are discussed. In May 1986 Japanese firms, under the direction of NHK, proposed their 'Hi-Vision' system as a global standard and some American firms supported this. The European members of the meeting feared they would be relegated to a second-rate position in this future technology; lobbyists created a united European block against the proposed standard, and succeeded in delaying the CCIR decision, by asking for some studies and for alternative proposals. These proposals were to be presented at the next meeting, i.e. four years later. A month later, in June 1986, the Eureka EU95 programme was launched as one of the first research programmes within the Eureka framework. The aim was to work on a European proposal for a global standard of HDTV; the budget was about 350 million dollars for the first phase and 500 million dollars for the second phase (after 1990). From then on, HDTV became *Highly Debated Television* in Europe, as it sometimes is referred to in the press.

The resulting tug of war around HDTV involved considerably more than interest politics around standards and markets. It is obvious that interest politics will take place because the economic and political interests connected to HDTV are immense. The potential world market is estimated in terms of hundreds of billions of dollars. Thus, the size of one's share of this market will have important consequences for other industrial activities, for the labour market, and for the relative strength of national economies. So when parties interested in the development try to gather support or try to promote this (future) commodity, this is not surprising. What is surprising, however, is the way interests get defined. It is only because HDTV is defined as promising that the interest politics can come into play. If HDTV were not seen as promising, nothing would be at stake.

The official *legitimation* of the proposed efforts uses an instrumental register, that 'the new is necessary because the present one is not good'. In the instrumental way of thinking the new technology must be seen as the solution for a problem - but what kind of problem? The problem of present television is not being formulated in terms of complaints - such as for example 'after two hours of watching TV one gets a headache' - but in terms of what is seen as technologically feasible. If the quality of the sound in our television is less than that of the best sound system available at the moment (the compact disc), this tends to be referred to as a *limitation*. A report that investigates the potentials of HDTV notes:

> The quality of sound of the present systems is not very good, especially compared with the quality of sound of compact discs (CD). Also the number of channels of the present systems is limited.[11]

The solution for such a problem is obvious. HDTV must (and will) deliver sound quality at the level of the compact disc. The same story goes for the size of the screen, the number of sound channels, the number of lines per image, the frequency of the scanning lines. All are less than what would be possible, or better, what should be possible, since the actual achievement of these specifications still requires much effort. Thus, the main shortcoming of the present system is that it is not HDTV. The present system is deficient exactly to the degree that it does not live up to the projected characteristics of HDTV. The present is measured by the yardstick of the technological promise, and found wanting. As a result, the development is self-justifying.

One explanation is that, as a consequence of their mandate, engineers, designers and other technologists are actively pursuing new possibilities; invention is part of their job. They will 'follow' the developments in their research area; that is, they must check what has been tried elsewhere, what presents itself as a new *option*, and what the implication would be for the present technologies and for market relations. At meetings in laboratories, in technical journals, and during conferences new options are presented and the community of technologists will consider whether these are 'promising'. The outcome, i.e. whether an option is accepted as a promise, operates at the collective level. All members of the community need not agree, but the outcome is often forceful, and often as an unintended result of interlocking activities.

In general, in technological development assessments of what is 'feasible' and what is 'obsolete' play an important role. Technologies are not replaced because they malfunction, but because another 'better' technology has become feasible. The present television system does not fail, nor did the black and white television of the 1950s and 1960s fail, or the good old typewriter, or the horse-cart. An 'old' technology does not fail in the sense a theory may be said to fail or a hypothesis that may turn out to be wrong:

> We are often misled by the cliché that technology proceeds by trial and *error*: abandoned technologies are in no sense erroneous (or failures or falsified). Watt steam engines, Bessemer converters, clipper ships or Fordson tractors presumably work as well as they ever did (Hamlin, 1992 p.529, emphasis in original).

An old technology does not fail, but is *abandoned*, and this is the reason why it does not function anymore, except perhaps in museums. They fade away like old soldiers. An old technology is obsolete, we say, it has been rendered out of date and is superseded. It is significant that the metaphor of a 'new generation' superseding the 'old' one is widespread. The notion of generation suggests that it is natural to replace it by a new one. The argument goes on by taking technological development itself as a proof that the old generation is outdated: black and white television is outdated by colour television, which in its turn will be rendered out of date by the introduction of HDTV; the typewriter has given up its place in favour of word processors and the horse-cart is succeeded by the motor car. The tautology is clear. In the world of action, however, tautologies can be important to give coherence to action.

As indicated, the mandate is a mixed blessing for technologists. On the one hand they enjoy some autonomy to define promises and to materialise them. But, once defined as promise, action is demanded. So, for technologists, self-justifying technologies such as HDTV appear as a two-headed monster. On the one hand, the development seems inescapable. HDTV is coming, whether we want it or not. Or as chairman Bögels of the Dutch Platform for HDTV (that is to prepare the introduction) put it 'The train is in motion. That train cannot be stopped any more'.[12] So ignoring the development is impossible; the train will continue without you.

But, on the other hand, leisurely leaning back and looking out of the window, knowing that the train cannot be stopped, is also impossible. On the contrary, hard work is necessary to keep things moving. The infrastructure has to be adequate; the knowledge level has to be kept up (Bögels: 'After all, through Eureka, we have been able to make up for our knowledge deficiencies. Therefore, we must now continue the European co-operation in HDTV'[13]); cable proprietors, programme writers and broadcasting companies must adapt to the demands of HDTV; new services that can make the system profitable must be designed and tried out; standards for the new system must be set down in law. If one of these - and the list is certainly not complete - fails, HDTV is delayed, and this is considered as a problem that must be solved as soon as possible. Bögels, for example, a year after his comparison of HDTV to an unstoppable train, worries about political hesitations regarding setting a world standard for the new system. He presses for clarity on this, for the decision has already been postponed once, and 'Another postponement would be quite bothersome, *because* this might delay the introduction of HDTV'.[14] In other words, as soon as the massive inevitability starts to show cracks, this must immediately be mended.

From Promise to Requirement

The resulting dynamics of this 'unstoppable train', and of self-justifying technology in general, is a conversion of promises into requirements. Once technical promises are shared they demand action, and appear as a necessity for technologists to develop, and for others to support them. At the same time, the options that are considered feasible and promising are translated into requirements, guidelines and specifications.

The study on the turbojet revolution by the historian of technology Edward Constant, is elucidating in this respect (1980). The study is a detailed account of how in the 1930s of this century, the propulsion of aeroplanes changed radically. Propeller-cum-piston engines were replaced by gas turbines. Constant frames this bit of technological history in terms of a *turbojet revolution*, referring to Thomas Kuhn's scientific revolutions. The case of gas turbines was not an actual anomaly causing the turbojet revolution, but a *presumptive* anomaly. The capacity of propeller-cum-piston engines appeared as an anomaly only in relation to the extrapolation of the gradual increase of speed of aeroplanes. Within that imagined context, which was generated by extrapolation, the piston engine failed, and a radical change seemed timely. 'When the gas turbine is the future of the aviation industry, then we must make it our task to develop it', is the argument. Indeed, Constant describes how a community of technologists was formed that defined their tasks and specified their goals in terms of the (non-existent) turbojet engine. Some made calculations of aerodynamics, while others searched for optimal turbine blades or studied materials. So the projection of the turbojet led to a task division and to specifications for the search processes.

The same can be observed in the case of HDTV, where the specificities of the promise serve as goals for R&D activities, such as signal processing, chips design, display and recording techniques. Some quotes from the OTA report (1990) prepared for the American Congress (emphasis added):

> HDTVs *must* process huge quantities of information at speeds approaching those of today's supercomputers in order to display a real-time, full-colour, high-definition video signal. HDTVs are able to do this at relatively low cost through the use of circuitry dedicated to specialised tasks (p.1).

> The much higher quality pictures of HDTV, however, *will require* the transmission of substantially more information than current TV systems [... requiring] the equivalent of five TV channels. To reduce this a number of tricks are used (p.52).

> HDTVs similarly place heavy requirements on memory technology. Access times needed for HDTV memory chips *must* be roughly 20 nanoseconds (ns) - 20 billionths of a second. Today's fastest DRAMs have typical access time of 60 to 80 ns (p.65).

> To truly appreciate HDTV, much larger high-resolution displays *are needed* than are generally available today (p.68).

However different these activities are, the technical promise sets their goals and co-ordinates the division of labour in terms of the overall objective of realising the promising technology, HDTV. The division of labour and the strategies are such that one can speak of a 'self-fulfilling prophecy'.

An example where the self-fulfilling character is formulated as a law is development of memory chips, in which all large microelectronic firms of Japan, Europe and the United States were involved. In 1964 Gordon E. Moore, research manager for Fairchild Semi-conductor, observed a regular doubling of the number of 'gates' (a measure for complexity), and claimed, by extrapolation, that this would continue. This prediction has come true so beautifully, that nowadays we speak of 'Moore's Law', as if it were a law of Nature. The validity of this law cannot be understood from the technical procedures by which the chips are made. The fact that the law holds so well is an effect of the way actors (in industry, in science and in government) judge their own and each others' accomplishments in comparison with what Moore's Law predicts. They direct their efforts towards achieving the predicted values. Laboratories evaluate and plan their efforts in terms of Moore's Law; when the specifications might not be met at the predicted moment, extra efforts are exerted. Firms base their investment decisions on the Law, for example the development of products that need chips with the predicted capacity - such as calculators or compact disc players. The government is willing to provide subsidy in order to avert the danger of not meeting the predicted value. Each actor exerts these efforts in order to measure up to the competition and to stay in the race. Moore's Law is the yardstick for the behaviour of chip producers and governments in Japan, the United States and Europe, and profiles their mutual dependence in the strategic game among these player (Van Lente and Rip, 1998). Moore's Law is not a law of nature, but a rule in the strategic game. As long as this strategic game is being played, Moore's Law will remain valid.

Conclusion

Technological futures are forceful. Once defined as a promise, action is required. This force can be understood as an outcome of language strategies and social arrangements, which, in their turn, are affected by long term historical changes. In particular, this chapter has examined the force of technical futures in three steps. First, I considered 'technological progress' as an ideograph. Support for concrete technological developments can be mobilised and legitimised by presenting them as instances of 'technological progress', in particular just because the content of an ideograph can vary considerably. The notion of ideograph enables the analyst to trace where technological progress in general is picked up, which actors use it as a resource and how it is realised eventually. The second step is the analysis of what I have labelled 'the mandate to technologists', which refers to the phenomenon of a societal division of labour of 'technologists' versus the rest of society. The interaction of technologists with their surroundings is regulated in terms of the general promise of technological progress. Their work, their opportunities and their obligations can be described as a mandate since they are relatively free to pursue their activities, but they must also be able to indicate that they do not neglect technological progress. If a promising technology becomes salient, they cannot really permit themselves to ignore this.

The final step in the analysis was to consider the implication for the dynamics of concrete technological developments. Due to the ideograph of technical progress and the mandate to technologists, technological promises function as a yardstick for the present and as signpost for the future. Thus, what starts as an *option* can be labelled as a technical *promise*, and may subsequently function as a *requirement* to be achieved, and a *necessity* for technologists to work on, and for others to support. This option-promise-requirement-necessity sequence does not imply that it is an autonomous socio-technical process. The transitions do not occur automatically, but are the result of actions and interactions of technologists, firms and governments. The transitions are a consequence of actors assessing what is 'feasible', what is 'obsolete', and what is 'necessary', and of their efforts following the assessments. Moreover, the transitions are reversible, in principle, and can be made undone - with increasing costs and work, though, because activities have become interlocked. The pressure to recoup sunk investment, for instance, increases. When after some time much has been invested in a promising technology, a detour or even a delay will encounter much resistance.

Yet, when technological futures find a ready ear, as they often do, the necessity of developing the technology may become so pressing that technologists, in a sense, are victims of their own promises. They cannot do anything else but work on that technology. Because the ideograph of 'technical progress' is self-sufficient and, thus, remains uncontested, technological futures appear as two-headed monsters. That is, they cannot be avoided because progress cannot be stopped, *and* they require efforts to be made because progress should not be stopped. In this sense, technologists cannot but help devoting themselves to the promising technology. Only to this extent is technology ever autonomous.

Notes

1. This is a revised version of 'The Cultural Space for Technological Promises' in my PhD thesis on *Promising Technology* (Van Lente 1993).
2. See also, Ellul (1967) and Mumford (1934).
3. Such a position is not unknown in sociology. It relates, for example, to Ann Swidler's (1986) treatment of cultural symbols as repertoires or tool-kits that individuals can select, instead of fixed patterns that exert power behind the back of individuals. According to the sociologist of culture Robert Wuthnow (1992), a *revolution* has taken place: 'It was once commonplace for theorists and researchers in the human sciences to speak of culture as if it consisted of deep norms and values, beliefs and orientations, predispositions and assumptions - all with an implicit stability and orderliness to which speech and action merely gave expression. Now ... speech and action themselves have risen to prominence ... values themselves are now said to be constructed in speech and action'. (Wuthnow 1992 p.1).
4. A similar point - of 'flexible words' having a force - has been made by Murray Edelman (1977) who in his *Words that succeed and policies that fail* indicates how for instance 'public opinion' is used as a strategy by different parties in the poverty debate in the United States: 'To define beliefs as public opinion is itself a way of creating opinion, for such a reference both defines the norm that should be democratically supported and reassures anxious people that authorities respond to popular views. In short, 'public opinion' is a symbol whether or not it is a fact.... In common with words like 'democracy' and 'justice', statements about 'public opinion' help marshal support for particular policies' (pp.49-50).
5. Many examples are given in Christopher Toumey's (1992) discussion of 'The Moral Character of Mad Scientists'. His point is normative: he warns against a dangerous anti-rationalist undercurrent in our culture: '...outside the circles of academic etiquette there is another kind of critique, a kind of Gothic subterranean reality, which reveals a visceral fear of science' (p.433). 'Stories of mad scientists, whether textual or cinematic, constitute an extremely effective antirationalist critique of science. They thrill their audiences by bringing together suspense, horror, violence, and heroism and by uniting those features · under the premise that most scientists are dangerous' (p.434).
6. David Baltimore, quoted in Jon Beckwith (1977 p. 234).
7. John Kasson (1977), in his *Civilizing the Machine: Technology and the Republican Values in America 1776-1900*, explicitly connects 'technical progress' and the status of those inventing technology: 'From the 1820s onward, ... Americans increasingly identified the progress of the nation with the progress of technology, and native inventors became the objects of a national cult ' (Kasson 1977 p. 41).
8. Ortega does not go into the question of the timing of this transition, since his main goal is to characterise 'technology' as a cultural phenomenon. It should be located in the beginning of the last century, with the foundation of engineering schools as an important ingredient. In the same period the meaning of the term 'technology' shifts from 'the study of skills and trades' to 'purposeful invention'. The introduction of the new meaning is often attributed to Jacob Bigelow (1787-1879), whose treatise on *Elements of Technology* appeared in 1829 (See Kasson, 1977; Staudenmaier, 1985).
9. Again, this has its parallels in many others domains. Sally Wyatt (this volume), for

instance, discerns a similar constituent role of the internet magazine *Wired* in the formation of an identity for an otherwise less clearly defined innovation network.

10. An extensive overview can be found in Brinkley 1997. See also OTA (1990), Slaa and Siderius (1990), Van Lente (1993) and the *HDTV Newsletter*.

11. Slaa and Siderius (1990, appendix V). The report continues: 'These shortcomings [of the present system] are due to the following choices:
- the low frequencies of the screen
- the choice of interlining in scanning
- the limited number of lines for each image ...
- the screen ratio of 4:3
- the limited possibilities for sound, in quality and quantity.'

12. *NRC Handelsblad*, 28 January 1989.
13. *NRC Handelsblad*, 28 January 1989.
14. *NRC Handelsblad*, 28 March 1990 (emphasis added).

References

Beckwith, J. (1977) Present and Future Houses, in *Research with Recombinant DNA, An Academy Forum*, 7-9 March, DC National Academy of Sciences, Washington.

Brinkley, J. (1997) *Defining Vision: The Battle for the Future of Television*, Harcourt Brace, New York.

Bush, V. (1945) *Science the Endless Frontier*, Washington D.C, National Science Foundation (reprinted 1960).

Callon, M. (1986) The Sociology of an Actor-Network The Case of the Electric Vehicle, in

Callon, M., Law, J. and Rip, A. (eds), *Mapping the Dynamics of Science and Technology*, The Macmillan Press Ltd, London.

Callon, M. (ed) (1999) *The Laws of the Market*, Blackwell Publishers, Oxford.

Constant, E. (1980) *The Origins of the Turbojet Revolution*, The Johns Hopkins University Press, Baltimore and London.

Hamlin, C. (1992) Reflexivity in Technology Studies Toward a Technology of Technology (and Science)?, *Social Studies of Science*, 22, 511-44.

Hill, S. (1988) *The Tragedy of Technology. Human Liberation versus Domination in the Late Twentieth Century*, Pluto Press, London.

Kasson, J.E. (1977) *Civilizing the Machine. Technology and the Republican Values in America 1776-1900*, Penguin Books, New York and Harmondsworth.

Landes, D. (1969) *The Unbound Prometheus*, Cambridge University Press, Cambridge.

Latour, B. (1987) *Science in Action. How to follow Scientists and Engineers through Society*, Harvard University Press, Cambridge, MA.

Marvin, C. (1988) *When Old Technologies Were New. Thinking About Electric Communication in the Late Nineteenth Century*, Oxford University Press, Oxford.

McGee, M.C. (1980) The Ideograph A Link Between Rhetoric and Ideology, *The Quarterly Journal of Speech*, 66 (Febr.), 1-16.

Ortega, Y. Gasset, J. (1962) Man the Technician, in idem, *History as a System*, Norton, New York, 87-164 (Original text from 1940).

OTA, (1990) *The Big Picture HDTV and High-Resolution Systems*, Washington, DC U.S., Government Printing Office. Report of the U.S. Congress, Office of Technology Assessment, OTA-BP-CIT-64.

Slaa, P. and Siderius H.P. (1990) Hoge Definitie Televisie. Een overzicht van de huidige stand van zaken met betrekking tot ontwikkeling en introductie,   SWOKA/VU,   Den Haag/Amsterdam.

Staudenmaier, J. M. (1985) *Technology's Storytellers. Reweaving the Human Fabric*, MIT Press, Cambridge, MA.

Toumey, C. (1992) The Moral Character of Mad Scientists A Cultural Critique of Science, *Science, Technology and Human Values*, 17, 411-37.

Van Lente, H. (1993) *Promising Technology: The Dynamics of Expectations in Technological Developments*, PhD Thesis, University of Twente, Enschede.

Van Lente, H. and Rip A.(1998) Expectations in Technological Developments: An Example of Prospective Structures to Be Filled in By Agency, in C. Disco and B.J.R. van der Meulen (eds) *Getting New Technologies Together. Studies in Making SocioTechnical Order,* Walter de Gruyter, Berlin, 203-229.

Van der Pot, J.H.J. (1985) *Die Bewertung der technischen Fortschritts. Eine systematische Uebersicht der Theorien*, Van Gorkum, Assen.

Winner, L. (1986) *The Whale and the Reactor. A Search for Limits in an Age of High Technology*, University of Chicago Press, Chicago.

Wuthnow, R. (1992) New direction in the study of cultural codes, in R. Wuthnow (ed), *The Vocabularities of Public Life*, Routledge, London, 1-18.

4 The Narrative Shaping of a Product Creation Process

J. JASPER DEUTEN AND ARIE RIP

The promise of modern biotechnology has driven investment in research and development (R&D), in new product development and in the continuing, even if precarious, success of small (and now larger) biotechnology R&D companies. The rhetorics and dynamics of promising technologies are not limited to modern biotechnology; they may well be constitutive of modern technology (Van Lente, 1993; see also this volume Chapter Three). Future worlds are sketched as a justification for investing in technological development. Different actors contend, and do this also by sketching their particular future worlds (see, for example, Hughes, 1983). In biotechnology, the arena of contestation has been expanded to include critical professionals, consumer and environmental groups, which are concerned about the possible impacts on environment and evolution, and about the risks of genetic modification to produce 'Frankenstein' foods. These are all public or semi-public arenas. In this chapter, we focus on a biotechnology firm, and in particular on the future worlds projected through its product creation processes - the product in this case being an industrial enzyme.

Within an industrial firm, such broader issues are often relegated to Public Relations Departments. And perhaps rightly so: innovation and new product development are difficult enough in their own right, and the people working on it should not be distracted from their main purpose. The broader contestation, and the rhetorics of risk and promise (Rip and Talma, 1998), are excluded to be able to get something *done*. Biotechnology firms have, by now, learned (and sometimes the hard way) that what they get done in this way may not be acceptable, and accepted, in society. There is a new receptivity to include broader societal considerations into the decisions guiding the product creation process.

While we applaud this change of hearts, we also suspect that it may be an add-on, tacked on to the regular management of innovation, for example

65

by inviting a spokesperson from an environmental group for a discussion session. Even when top management is serious about these issues, such interactions may remain symbolic because there is insufficient understanding of the way promises and warnings function in product creation processes. This is why we have bracketed out these wider concerns, and focused on what we call the 'narrative shaping' of product creation processes. Only if this aspect of product development is understood and taken into account will the wider interactions be productive.

Thus, we will focus on the firm-internal processes, and highlight narrative dynamics that serve to constitute an actor's future. We will do so in general terms, by criticizing modernist stories about successful innovation. We will then present the mosaic of stories which compose the development of the industrial enzyme Gammese (the name is fictional), and continue to identify generalizable patterns. Contestation is visible, whilst remaining subdued. But the dynamics of future worlds, and their inclusion in ongoing and interacting narratives, may well be a general rather than unique pattern of future oriented organisation. On the basis of our case study, we hope to demonstrate the way in which this analysis both highlights the constitutive character of narrative whilst also critically contesting the accounts of actors who sought to explain to us how their present future came about.

Innovation Journeys and Narrative Analysis

In retrospect, the story of a successful innovation is often told in a linear way, with the first plans leading 'naturally' to the eventual outcomes. In these accounts the eventual achievement functions as a goal to be reached from the beginning, and is realized in a number of steps, the stages of a journey along the path that had been visible from the beginning. Actual processes, however, are much less linear than these retrospective accounts suggest. The metaphor of an innovation journey, with its contingencies, its setbacks and its detours, captures the real-life complexities of product creation processes much better than rational-control views (Van de Ven et al., 1989). Therefore, linear accounts will often be a simplification and distortion of a more complex process. So, is there something to be learned about the narrative shaping of a new product? Something can be learned if one realizes that accounts are produced all the time, not just after the journey has ended. There is a variety of accounts: formal and informal, technical and social, strategic and operational, for internal and for external purposes. These accounts are linked and build on each other. So one can inquire how such accounts evolve along the journey, and why they can become more linear over time. Linearity turns out to be an outcome (albeit a precarious one) of interacting narratives, rather than a necessary feature of product development. During the journey a certain thrust and directionality can develop.

Besides linearity and thrust, interacting stories have other (narrative) effects. We use the notion of an emerging 'narrative infrastructure' to analyse the overall effect of interacting stories. Analytically, the important point about infrastructures is that they help to explain how coherence and linearity can emerge in multi-actor, multi-level processes, without any one actor specifically being responsible for it. Product creation processes are one example of emerging coherence. They might well be a specific genre with a typical form of narrative infrastructure.

We have reconstructed the actual innovation journey in the case of an enzyme, to be used as an additive in animal feed. This innovation project had its setbacks and detours, but was successful in the end. Before telling the story of how scenarios and other narratives shaped the innovation journey and built up an overall thrust and linearity, we have to develop the narrative approach a bit further, and give an account of our method of data collection and analysis (particularly, the issue of retrospective accounts).

Narrative Infrastructure

First we have to clarify what we mean by 'narrative' and how it relates to future-oriented action. Narrative occurs in interaction, it informs and shapes action, and makes action into something memorable. Narrative and (the need for) action are closely connected. Our suggestion is that agency appears only in and through narrative. In other words, narrative is constitutive of agency - instead of the other way around as is often supposed. For example, a promising story of modern biotechnology drives investments in R&D and new product development. Or, more specifically, in product creation processes, a project team is constituted and acquires space to work on product development as an effect of prospective stories: 'selling' a lead for a new product and portraying itself as the 'hero' who will be able to achieve the desired new state. Agency materializes in this way, also literally (Law, 1994; Van Lente and Rip, 1998), and an overall thrust is gradually built up. The key point for our purpose is how the contingencies, even chaos, of ongoing interactions are shown to acquire a shape, in fact a variety of shapes, through the stories told, at the time and afterwards.

We use narrative in a broad sense. The actual telling of stories, whether prospective or retrospective, whether terse or elaborated, is only one part of our use of narrative. Narrative in the broad sense takes the material setting and the situation into account. The staging of the text of the story is an essential element of the 'story'. So, there is more to 'story' than words and a receptive listener/reader. Rather the reverse is the case: the 'story' is produced by the setting, in the broad sense, and the actions and interactions played out in and with it. The actors are not just tellers/authors or listeners/readers, but they become characters in the overall mosaic of 'stories'. The actual stories they tell are only one of the elements contributing to the evolving 'story' or mosaic of 'stories'.

It is here that our notion of narrative infrastructure comes in. On a first, and superficial, level there are terse and elaborate stories told by the actors (Boje, 1991). The teller of a story has a listener who will respond and become the author of a further story, building on, adapting and/or contrasting the earlier story - always in the broad sense, including material and social aspects. This turning of the narrative tables in ongoing interactions creates a multi-authored and always heterogeneous mosaic of stories. Sometimes, one master story evolves. What always happens is that some of the narrative building blocks continue to be taken up, become accepted ingredients, and because of their being accepted, orient further action and interaction in the setting (and across its boundaries). The building blocks and their linkages constitute a narrative infrastructure, which enables as well as constrains. When a narrative infrastructure evolves out of the stories, actions and interactions of the actors involved, actors become characters that cannot easily change their identity and role by their own initiative.

Product creation processes can be seen as one genre of overall 'story' in which novelty and uncertainty are important aspects of the setting. Actors, in fact, speak of the 'story' of the creation of the compact disc, or the personal computer. In our case study, of a biotechnology firm developing a new industrial enzyme, Gammese, actors spoke easily of the 'Gammese story', and could compare and contrast it with other such 'stories'.

Narrative Analysis

Narrative analysis of such broader 'stories' draws specifically on the narrative analysis of texts. For example, there is sequentiality (or constraints of the past, or increasing irreversibility), not just as a matter of choices being made by actors, sunk investments etc., but through an evolving narrative. The reader-author collusion (predicated on a shared culture) imposes constraints on what can be said, and similarly, the triangle of actors, setting, and narrative infrastructure enables and constrains action and interaction. In texts, for example, if character X has been introduced as male, it becomes almost impossible to let him become pregnant. In organizational life, there are role expectations and specific cultural repertoires. And there are problem definitions and typifications, including views of what kind of product it is that must be created (which shapes the innovation journey) and views of what various strategic partners mean for the product creation process (which foreclose other options).

This type of analysis is necessary to trace the development of a certain thrust over time in the multi-actor, multi-level product creation process. In addition, typifications develop which become part of the narrative infrastructure and constitute the building blocks for an eventual master story. Successful product creation processes have 'heroes' and 'helpers' (and failures may have 'tragic heroes'), and are thus amenable to Greimasian semiotic analysis of actants in a story (Greimas, 1987). We shall follow their approach only loosely, however, because we are not limited to a written text, and some of the distinctions and figures introduced by Greimas lose their force when the story is multi-authored and interactive. In making this move from textual semiotics to social semiotics, we follow actor-network theory (Latour, 1984; Callon, Law and Rip, 1986). While some concepts of actor-network theory, like enrolment and translation, as well as some of the case studies (Callon, 1986b), suggest entrepreneurial voluntarism (and have been criticized for that), it is the interest in emerging irreversibilities (Callon, 1991 and 1992) and infrastructures (Latour, 1984, cf. also Van Lente, 1993 p.212-223) which is important here.

An example of narrative analysis of the thrust of a project and its evolving story is Van Lente's (1993) study of a failed innovation (see also Chapter Three of this volume). Particularly interesting for our purpose is his detailed tracing of prospective stories and their interaction, reinforced by assertions that the 'right' thing is being done. It then becomes difficult to say that a project should be stopped, and if such a proposal is made, it disorganizes and embarrasses actors - because their narrative infrastructure does not support them anymore. The case study concerned an innovation project aiming to develop a new isolating material, Tenax, important in the world of high-voltage transmission of electric energy, and being pushed on the basis of expectations about its potential performance. Researchers, managers and members of the board of directors told stories of progress (actual and expected) for a number of years - and rightly so, in spite of difficulties, including the practicalities of producing high-voltage cables. The effort to maintain progress became too high, however, and in the space of one month assessments were turned around. To the surprise of the Board of Directors and some external allies, the project collapsed. As if it were a house of cards - and indeed, it was a house of cards, because its strength resided in stories that had to come true. Interestingly, the theory about the electric performance of the new material, at first presented as a robust resource, now became 'just a theory', and the research institute KEMA propounding this theory was transposed from an ally to the scapegoat, the source of failure.

Our own case study is one of a successful project, but one which was on the brink of collapse a number of times. In other words, success or collapse are not the main distinguishing variables. The underlying dynamics are more important - which leads to the question of how to reconstruct them.

Data Collection

Reconstructing narrative in a retrospective case study is beset with difficulties. Sometimes, there is enough documentation on the early stages to get a view of the variety and the contingencies at the time, independent of the reconstructions by interviewees. We were fortunate in having access to all the project team files, the minutes, notes and letters, and official documents. These data were used to reconstruct processes and interactions (Deuten, 1994), and as an input for the interviews.

A successful product creation process also makes alternatives invisible, and contingencies along the innovation journey are then seen as noise, or perhaps occasions in which the prowess of the victorious hero was shown. Trying, in interviews, to get behind such actor's reconstructions will then seem to undermine their victory. Even when interviewees do not feel threatened, there is still the effect of outcomes being known, so that events, choices and actions at earlier stages will be presented as part of a development leading toward this outcome. Interviewees will automatically introduce characterizations in terms of 'right' or 'wrong' (just as watching a play or a movie where 'seeing' the storyline enables us to identify heroes and villains quickly).

One way to obviate such reconstructions after the event is to ask the actor to time-travel, and think back to the earlier situation. Documentary data and imaginative stimuli by the interviewer help him to remember the uncertainties and contingencies that were lived through, and get him to tell about them. (We say 'him' because most of the actors in our case study, and all our interviewees were male). In this way, one can, on occasion, also see how contingencies were reduced and linearity was created by introducing narratives with a certain plot which, through being told and linked to the stories of others, became true.

Another entrance point is provided by prospective stories told at the time, from expectations and the organisation of agendas into scenarios for uses of the product and the market assessments at an early stage of the product innovation process. Action is shaped by such stories. Documents of this kind from the project file were discussed in the interviews to find out about their setting and the role they played.

Our two main interviewees, Orlans and Bentrom (these are fictive names), were in a position to see themselves as agents, as persons who made a difference. Orlans was head of non-division R&D and responsible for pushing the project in its early phases, Bentrom was leader of the project team. Both were also natural narrators, and realized how they had been using stories to further their ends. Our method of 'time travel', putting them back in situations where we knew shifts in context or content had occurred (based on the detailed chronology we had set up using archival materials), worked well with them. We asked them to describe, not to justify (or condemn), and obtained materials showing a mix of contingency and purpose, reflecting uncertain responses to setbacks, and exemplifying how they tried to create agency and linearity. (Of course, all materials from interviews are joint constructions by interviewee and interviewer. But the construction is not arbitrary, so the result tells us something.) We also heard about the stories they consciously told as management tools, to team members and to other levels in the company (see Deuten, 1994).

In the next section, we shall present our data as indicating the evolving mosaic of stories which constituted the Gemmase project. Our presentation implies a meta-story in which a narrative infrastructure emerges, and we shall highlight the meta-story in the subsequent section.

The Gemmase Project as a Mosaic of Stories

Reduction of complexity and uncertainty is important to get a project started at all, but can, of necessity, be only tentative at that time. Management decisions and the resolve to get something going are taken, in retrospect, as the beginning of a project, but are themselves the outcomes of earlier and less clear processes. In the case of the Gemmase project, within extradivisional R&D an old idea about using the enzyme Gemmase as a feed additive (to improve the uptake of phosphates) was being reconsidered in the early 1980s. A contact person from the feed sector had told them that there might be a market for such a product. Because of the progress made in recombinant DNA technology, the production of this enzyme might turn out to be cost-effective.

Our interviewees stressed (in line with received views in innovation management literature) that a promising idea or a 'lead' must be transformed in a clear concept of the technology, the functions, the applications, and with expectations about cost of production and potential market - all this at a stage when very little can be said with certainty. Otherwise, they said, there is no orientation of action, nor can one convince others about the value of the idea. But making a clear concept is not just a matter of listing arguments. Orlans arranged them in a story about a world where Gammese would play a role: as an essential ingredient of animal feed (reducing costs for farmers as well as reducing the environmental burden of intensive farming) and as a key element in the strategic portfolio of his company. In other words, a trustworthy start-out story is essential in the early phase of a project. The start-out story is like a scenario, made robust through linkages with scientific, technical, economic and strategic elements, as well as the credibility of its authors (for Brown, in Chapter Five of this volume, Dolly the cloned ewe is a good illustration of a poorly articulated start-out story whereby the utility and value of the event was too ill-defined to protect it from wider social critique).

Orlans, in fact, insisted on the importance, in product innovation processes, that 'there is somebody with vision and credibility, who convinces the others that this must be accepted' (note the use of 'must'). He was such a person, and without him, he said, the project would not have taken off. Orlans actually spoke of himself as a 'product champion': representing the product-to-be to the world, but with the connotation of being a fighter who turns setbacks into challenges. Such a typification is an easily available role/identity in the repertoire of management culture (and in the management literature). In his case, he presented the promise of Gammese to other divisions, staff and the Board of Directors, realizing the multi-actor and multi-level dynamics involved and playing on them.

The start-out story was reinforced and convinced the Board of Directors. Part of the R&D budget was made available by the middle of the 1980s, and a small project team was constituted. A limited *in vivo* test with a known type of Gammese was done which performed very well inside animals. However, within the company there was some resistance to the project: would there really be a market for industrially produced enzymes in animal feed? There was no way of telling directly. A pessimistic as well as an optimistic scenario existed about the future of Gammese, both of them diffuse. The project team saw its task as making the positive scenario come true. An important step had been to involve a Working Party on Digestibility of Phosphates (*Werkgroep Fosfor-Verteerbaarheid*) of the Community Board on Feed (*Productschap voor Diervoeders*). The company needed the expertise collected in the Working Party, and together, they made a detailed planning of the steps in the development and first applications of Gammese. The diffuse scenario became specified, and it was co-authored by credible actors. The content and context of the project plan convinced the Board of Directors, and the Project Team could continue and expand. Two things happened at the same time (and are in fact two sides of the same coin): commitment and resources created a protected space for the project plan to be realized, and the Project Team became a unitary agent responsible for progress, and thus for the necessary repair work.

The project plan is an important element. It is a prospective story, setting out stages of the innovation journey. Since it is used in communication with higher levels, it is also an account before the fact, and the project team will be held accountable for deviations. The project team has to use the plan as a road map, even while realizing that the road is not there yet, and contingencies have to be faced. Finally, the project plan also allocates roles and tasks internally, and specifies linkages with external actors (within the company and outside it). It is a stylized story, with various characters and a (minimal) plot.

The project team would check against the milestones in the project plan, and work harder if these threatened not to be achieved. When such efforts failed, one had to have a good story to tell the Board of Directors. Repair work, in the small and in the large, was structured by the need to follow the plan and so to stay on course, rather than only by the need to solve concrete problems.

The relationship of the Project Team with the Board of Directors was ambiguous. The regular reporting to the Board of Directors, as well as the reporting in incidental interactions with them, has a double function: on the one hand, sharing information within the company, in particular with higher management, and on the other hand a project team saying to its sponsor we're doing (reasonably) well, please continue supporting us. It is a balancing act, as Bentrom experienced it:

> It is important to communicate uncertainties to higher management, although you have to be careful there as well, in my experience. ... You should prepare them so as not to have to surprise them later. On the other hand, if you indicate too many uncertainties, they say 'this won't come to anything, this guy is so uncertain'. Or in a less personal vein, You have to steer clear from various dangerous rocks. For one thing, you should not raise exaggerated expectations. For another, you should not paint too sombre a picture, otherwise they'll scrap the project.

Clearly, this is a dialectics of promise (Van Lente, 1992). In the case of Gammese, the dialectics could profit from widely-shared background expectations about the importance of enzymes, about markets, about regulation, so that Orlans could craft a convincing story and keep the project on its course. But circumstances could change: the relationship with the Board of Directors came under pressure in the late 1980s, when the company went through a process of strategic re-orientation. The company wanted to go back to its core competences. Enzyme production definitely belonged to the core, but capturing large slices of agricultural markets did not (even though the company had been trying to expand in this direction). Gemmase had to be repositioned to keep its support. Orlans and Bentrom successfully shed the connotation of Gammese as a commodity in the agricultural market, and convinced the Board of Directors that the company still had a role to play in this market, supplying Gammese as a specialty. The Commodity Board and animal feed firms were mobilized to support this claim.

The new story was further strengthened by emphasizing the environmental advantages of the product. Apart from the substance of the argument, there were also PR considerations. Not just for the product itself, but for the image of the company as a biotech company in a time when societal acceptability of biotechnological products was an issue. Bentrom:

> For some other enzymes produced by the company it was difficult to explain whether there was a benefit to the consumer. So it was noted that it was useful to have a product that is easier to explain. But it was not developed for that reason, of course. This was an additional advantage.

At the level of the company, the Gammese project helped to tell a story about the positive role of biotechnology in society. In the annual reports, the project was regularly brought up as a good example of the contribution of biotechnology to reduce environmental problems.

This turned out to be a mixed blessing for the Project Team. As early as 1987, the Board of Directors announced to the press that the company was working on Gammese. This was four years before the planned date of introduction on the market. The Board of Directors probably did so because it could score in the media with this environmentally friendly product (the fact that it would reduce phosphate burdens in agriculture was emphasized). In Bentrom's experience, this created an enormous pressure on their project.

> As far as I was concerned, there was no need to do such a thing. ... On the other hand, the advantage is that the company commits itself publicly to this project, so they can't stop it easily anymore.

The registration of Gammese was another problem that needed to be tackled, and where many actors at different levels were involved. At the time there was no relevant regulation in the Netherlands or at the level of the European Union yet, so the fate of Gammese was uncertain. Informal interaction of Orlans and others with officials of the Ministry of

Agriculture indicated that there was a possibility of ad-hoc admission. In Orlans's words:

> This registration question was of course a difficult business. The Netherlands would have to risk its neck in advance of an eventual EU regulation, and defend this in Brussels. [The Department of] Agriculture has had difficulty in doing that . ... We needed Agriculture. On the other hand, it was clear to us from the beginning that Agriculture needed [Gammese] [because it would help them solve environmental problems in Dutch agriculture]. ... We have been active politically, put forward our story there. ... So a story had been established of [Gammese] being an interesting product.

A Director-General in the Department of Agriculture found the promise of Gammese so interesting that he arranged (perhaps after some prodding from the company) that the Minister would come and visit, and hear the Gammese story from the company itself.

> So we had the whole club visiting us [the project]: the Minister and a lot of high-level officials of the Ministry. Our Board of Directors was there - that was a good thing for us, naturally - and then we told, in all its splendour, the whole story of what we thought was the role [Gammese] could play in the Netherlands, what with the environment and so on, and how far we were now with production. How we expected to have everything ready shortly, but that we needed approval, and what Agriculture was doing about this. But really, in other words, by showing off this whole story again, there was no way back. In this way, we also supported those people from Agriculture who were working on the approval, saying as it were: this must happen now, isn't it? All the big men were there, so if the people would encounter resistance, they could always say that their bosses had heard that it had go through. All that helped.

Telling the story to the Minister of Agriculture, externalizing it as it were, created commitments internally, with the Board of Directors, with the Project Team. And Orlans and Bentrom realized this, and exploited it.

Public acceptability was also a matter of concern for the project team. Public acceptability is important for every product nowadays, and especially if it is biotechnological (Deuten, Rip and Jelsma, 1997; Jelsma and Rip, 1995). Spokespersons for public acceptability therefore are important actors to the company. In this case, the company had to convince a Consumers' Platform on Biotechnology that Gammese was important for the consumer, and that it was safe. The environmental advantage of Gammese played a key role in the stories told to the Platform.

These interactions were actually part of a longer process, in which the company had been anticipating issues of acceptability and trying to avoid problems. Orlans explained this as follows:

> I have to add that this product, [Gammese], was not such a difficult product in this respect. In genetic modification, there are gradations from homologue to heterologue modification, and here everything was quite simple [because it was a homologue, i.e. less chance of unexpected effects], so we didn't have too complex things to do [for registration]. Also, we hadn't used markers or other things which could raise discussion. So we were on the safe side in this acceptability issue.
> [Deuten: Did you do all this intentionally?] We paid a lot of attention to it, from the beginning. Like let's not do it this way, because it will create a lot of problems for us. [Deuten: Were there negative experiences in earlier projects on these points?] Yes, we even had a kind of strategy in the company to build up acceptance very gradually, and preferably by starting with 'safe' ventures. So not go out and challenge the world, that would be too risky. [Gemmase] fitted perfectly in this strategy, otherwise we might not even have started the project. ... Of course, we had some experience with other projects. ... So you can choose the right directions. And we profited, of course, from the great advantage of the product being environmentally friendly.

Narrative Reduction of Complexity - And Its Risks

In the project planning a series of activities was formulated. First, the best Gammese had to be found in an extensive screening programme. Second, on the basis of the amino-acid sequence of the selected enzyme the DNA of the micro-organism had to be cloned. Third, a host had to be selected in which a DNA construct for over-expression had to be implemented. Finally, the production process had to be optimalized. Meanwhile, application tests had to be done and a formulation of the end-product had to be developed. The planning schedule was tight, and the different activities had to be managed in a parallel way. Delays in one line of activities would cause delays in another line of activities. During the project smaller and bigger problems and delays occurred. We shall give two examples of how management dealt with these uncertainties.

A major setback, at first not recognized for what it was, was the degradation of the enzyme when the feed with which it was mixed was pelletized. The project planning came under serious pressure. A series of earlier measurements of thermal resistance of Gammese had been quite encouraging, but now, in another set-up for making pellets, the enzyme degraded. When asked about it in one of our interviews, Bentrom said:

> I think we did not want to believe it at first. [Deuten: You thought it was a measurement error?] Yes, because we had shown a number of times that pelletization resistance was good. Then you don't let yourself be thrown off balance by one experiment which indicates that thermal resistance isn't as good as you thought. So we said, let's do another experiment. As yet, there's no reason to completely change course in the project. [Reflecting:] We absolutely refused it. That is a bit of denying reality. But what if you get good results twice with an enzyme, and bad results the third time, what do you do?

There was the psychological element of having lived within the framework of a story, and not wanting to give it up, since it would mean losing your road map (Wagenaar, 1997). There was also an effort at checking the 'reality'. At the time, it is not clear whether thermal resistance might indeed change in different circumstances, or whether one might perhaps control circumstances so as to minimize degradation of the enzyme. It is only after repeated attempts and assessment of their outcomes that one decides whether to 'change course' or not. During those attempts, the original story and road map remain the guideline. Fortunately, the problem with pelletizing was gradually clarified, and other ways of adding and mixing the enzyme (originally seen as less relevant) were taken up successfully. In the case of Tenax referred to already (Van Lente 1993), things didn't turn out so well. In this case the course was changed, for some quite unexpectedly, and the story was adjusted. In both cases we see how narratives create inertia for a project team in a protected niche. For actors, such an attitude of trying to stick to the original plan can be viewed positively as tenacious, seeing setbacks as a challenge to the 'purpose', but also negatively, as reduced ability to respond to changes.

Elsewhere (Deuten, Rip and Jelsma, 1997), we have analyzed this part of the case history as deriving from an early alliance with one selected lead user instead of a broader range of users, with whom the tests were conducted. After successful conclusion of these tests, the number of try-outs with other users was expanded, and it turned out that in their set-up the enzyme degraded. The dilemma for management is that early alliances are necessary, but clearly also a risk, if there are specificities (which one does not know beforehand). Although we do not have quotes from the interviews to this extent, we suggest that the Project Team was using a story line in which their early user had become typified as 'the' user, sufficient to represent all relevant users. In our second round of analysis, we shall indicate such a typification by writing THE USER, in capital letters to emphasize its generality. Here, the point is that typification entailed that inquiries about specificities were deemed unnecessary and other users were moved to the background. This is a general feature of typification, and we will come back to it in the next section.

After the near catastrophe of the product degrading under regular conditions of use, the Project Team tried to work with more than one option - as it were creating alternative scenarios which could be taken up in case the main road map threatened to destroy the prospects of the project. Besides this particular way of reducing, or at least handling, uncertainty and contingency, other ways were visible from the beginning. Schemes and planning were important to reduce complexity on paper, hopefully becoming self-fulfilling prophecies. Experts of various kinds were consulted not just to solve a problem, but also to be aware of possible problems.

Another example of narrative reduction of complexity, which oriented (and thus constrained) action for some time, is the alliance forged with a carefully selected foreign firm, well located in the markets of animal feed additives and expected to be knowledgeable about formulation technologies and about registration procedures. While the Project Team and the Board of Directors had put high hopes on this alliance, the specific expertise of the alliance partner appeared to be of little help in this case. The Project Team had created a character in their story of the product development process, the ally, to play an important supporting role. In other words, they had made a typification, THE ALLY. It took quite some time before they could believe that this partner did not avail of superior know-how for these specific enzyme formulation problems.

When looking back on this episode, Bentrom and Orlans still find it necessary to argue that there had been good reasons for the alliance, and/or that they could not be blamed for not checking more carefully. Clearly, there is a conflict between the dynamics of evolving accounts at the time (which can be understood narratively), and the need to present a consistent retrospective account now (which is narratively necessary, because the project turned out to be successful).

Bentrom and Orlans explicitly used stories to manage the project team: for team building, to make sense of unexpected events, or to motivate team members. Bentrom's stories were like the external scenarios constructed at an earlier stage to convince others, and in particular the Board of Directors, that they should support a project to develop Gammese. The difference is that he now uses events, views and stories from the outside to persuade his own team members of the importance of the Gammese project.

The comparison shows that narrative plays a role in the transition from project to environment, as well as the other way around. The thrust developing in and through the project derives from the linkages across levels and their precarious stabilization. Telling the project story to third parties, elsewhere in the company, possible external allies, and audiences to be appeased, leads to reciprocal expectations and commitments, whether it is done for substantial or tactical reasons. An author writing a fictional text is constrained by the features of his characters and plot, in relation to the author-reader collusion he wants to maintain. In the 'genre' of product creation processes, there is no single author, and no master text being written. But there is a similar reduction of possibilities (and thus of complexity and uncertainty) which enables the various actors to be productive, while at the same time constraining them in certain directions. Phrased in this way it is clear that this is a matter of narrative infrastructure.

Telling Yourself Forward, and Telling the Product Creation Process Forward

A certain thrust developed over time in the product creation process of Gammese-to-be. The narrative infrastructure that emerged shaped action and interaction, and helped to create overall patterns in the mosaic of stories so that finally there emerged the Gemmase story. It must then be possible to rewrite the case history in terms of characters and (evolving) plots, and so bring out its narrative character (in the broad sense). This will support, by demonstration, our general contention about the narrative character of the reduction of complexity and uncertainty and the building up of linearity and a thrust.

Characters in the Gemmase story, typified as 'hero' or 'ally,' and phrases like 'telling yourself forward,' are used as semiotic categories (in the broad sense, which we denoted as social semiotic). That is, they are not descriptions (in the modernist vein), but indications of plot and character as these emerge - but with strong implications for subsequent actions and interactions.

The stories told by Orlans and Bentrom to the team position them as a Gideon's band. They are the heroes who have to make the promise of Gammese come true. Institutional-memory stories support this effect by reducing uncertainties: we have had this problem before, but if we put in enough effort we can solve it. Stories about the importance of Gammese in the wider world have an ambiguous character: the project team leader uses them to motivate his team, but in doing so also has to set up Gammese as the hero in a story in which environmental problems are solved. A similar ambiguity is visible in the stories for the Board of Directors, where the Project Team works for its survival by positioning Gammese as the hero which solves environmental problems as well as public acceptability problems. In the interaction with interest groups, the only hero is Gammese.

This may be a general pattern, which implies that management by story telling should be located in a broader context in which resources and allies are mobilized and barriers are overcome by versions of the story that is used inside. Management by story telling, influencing sense making of team members, is then not independent of the links in those stories with the wider world.

Adding the links between the work unit and other levels of the organization, and with the wider world, the setting is recognized as part of the narrative. Thus, one can understand how the structure of the overall narrative reflects the telling of oneself (one's collective self) forward. This is particularly visible in external interactions: the internal interactions and narratives are black-boxed, and the black box is labeled with the intended product of the work ('we are the Gemmase Project') - while the product itself ('Gemmase') then becomes the main character in the external stories.

How the Project Team Became Part of Its Own Story

The start-up story sketched a future world in which the product to be developed turns out to be successful and helps the firm as well as customers/users, and it identifies a core group, the Project Team to be, as the character that must be supported. Roles are specified for various characters, who can become co-authors if they are willing to go along - which they may refuse. Such role specification and enrolling has been analysed before, for instance the electric-vehicle world projected by Electricité de France in the 1970s (Callon, 1986a). In this case, Renault was enrolled at first, but then stepped out of this world, which hastened its breakdown.

At first, the Board of Directors is a key character, an obligatory passage point because of its authority and power over resources. When the Board goes along, a protected niche is created for product development. The scenario for a future world has to be realized by the Project Team and its allies, and so a purpose is created at the same time. The purpose contains an element of general motivation, but also a story, an evolving project plan that functions as a stylized narrative guiding the various characters. Realizing the project plan creates agency: the Project Team will make a difference, at the same time as it will put Gammese on the map.

Orlans and Bentrom often positioned themselves as independent agents, enroling others, mobilizing resources to their own purpose, and framing and telling their stories to that effect. But telling a story in which you are a character yourself creates constraints as well: You become a character with a specified role in the subsequent stories of the listener/reader and you cannot permit yourself too much deviation from the expectations connected with from this role.

The Project Team positions itself as rising to the challenges of the innovation journey, and so cannot shift tack with respect to its plans and promises without losing its identity. This effect is reinforced by the need to tell, and continue to tell, stories to the outside. If these stories are accepted, the Project Team is now also a character in the stories of others, and cannot free itself from the obligations these bring with them without losing credibility or otherwise dropping out of the fabric of intersecting narratives it had been contributing to for its own purpose. The burden this creates may eventually become too high and the Project Team might give up - as happened in the TENAX case mentioned earlier, where the Project Team suddenly reversed on its promising stories, to the surprise of its Board of Directors and some of its outside allies (Van Lente, 1993).

While the Project Team is the central character and has to confront the challenges, it is not alone in its heroic task. In narrative terms: there are allies and subsidiary heroes. Its relation with the Board of Directors is ambivalent: as a benevolent sponsor the Board is an ally, but it is also a threat since it can withhold authorization and resources. The Project Team reports to the Board, and makes sure it shows how it follows the project plan, or else has good reasons to deviate from it.

Relevant actors become characters in the overall story. The lead user at whose plant tests would be conducted on formulation, and in particular the behaviour of Gammese during pelletization, becomes THE USER. This is a typification which blackboxes and thus obliterates the variety of circumstances of application. Similarly, the German firm with hopefully complementary expertise becomes THE ALLY. In all cases, the Project Team assumed authorial discretion to locate the character (including the human and non-human actors it contained) as it saw fit - and was unpleasantly surprised when the character went its own way.

Non-human actors participate in the narrative in the same way. Gemmase-to-be is part of the cast from the very beginning. Genes of Aspergillus and the possibility of modifying them in particular ways turn out to play a role in acceptability of the process. Properties of the enzyme are translated into functionalities, cost-effective production in the lab and then upscaling - these are part of the standard story of a product development process, and the non-human actors are assumed to accommodate to the roles assigned to them. Again, rather than allies and subsidiary heroes, they may turn out to be untrustworthy, confusing or even act as opponents in a battle that the Project Team might not win.

The Product Triumphant

Specific to narratives of product creation processes is the presence of what we call, for want of a better term, dual heroes. In the start-up story, a promising scenario about a world with Gammese to-be-developed allowed resource mobilization and the creation of a protected space for a project team with a purpose. The Project Team is the hero, but to continue its quest, it has to tell stories about their eventual product: how it will become profitable, how it will help the company present biotechnology as really useful for society, how it will support agricultural authorities in overcoming waste problems, etc., etc. Such stories are necessary, but derive their power from the setting and the interactions played out in it. A narrative infrastructure emerges in which another hero is born: Gammese itself, which will stand triumphant in the end. The Project Team, because of its own success, will become invisible.

We suggest that this shift from the innovator to the innovation as hero will occur in every product creation process, and necessarily so because the attempt to move forward on the innovation journey, involves inevitably the emergence of a narrative infrastructure which has the product to-be-developed as the main character. In isolated stories, told on particular occasions, one or the other hero will get the limelight. When the innovation project is seen as an evolving narrative, the complexities of the plot reflect the criss-crossing linkages between actors trying to position others, and being positioned by them. Because their shared reference point is the product to-be-developed, this will take on a narrative role of its own. When the innovation is successful, it will eclipse the agent which prepared its way.

The converse happens as well, as in the case of Aramis, a failed project for new subway vehicles and guidance systems, described by Latour (1992). The Aramis story is the tragic version of the 'product triumphant' plot. The innovation fails, and Aramis disappears as a character. In Latour's story, he fleetingly appears to Latour's alter ego, asking why he was not allowed to come to life, and accusing the alter ego of faintheartedness.

Reflections

We have demonstrated that product creation processes can usefully be studied with a narrative approach. We have shown how complexity and uncertainty is reduced, and presented as reduced, in accounts building on each other. We let some of the actors speak, while locating them in processes in which an overall thrust was built up at the price of constraints, in which problems were encountered partly because of the way the Gammese-story had been shaping up, and where a new hero was born precariously.

We also attempted to reduce the complexities of plots and characters emerging in this way, to make them intelligible and applicable to other product creation processes. For example, the identification of a story about how it all began, is itself an origin story, a projection - and thus a meta-story - on the complex and contingent streams of events and interactions at the time, which attributes originating force to some actions and interactions by selectively highlighting them. Such a meta-story feeds into another narrative infrastructure, which enables and constrains the discussion of the nature of product creation processes.

Our rewriting the product development process of Gammese enhances understanding, but also unsettles actors. When Orlans and Bentrom read our analysis, they recognized the points we made as real and valuable - but also felt slightly uncomfortable being positioned as characters in a story, and seeing their own modernist terminology between quotes. Managers typically write (i.e. produce texts and stories) in a modernist vein, assuming their own agency, and assuming readers who will follow them in their exposition, and who can be routed and re-routed. If they recognize the possibility of another genre, that of developing an interactive narration in which they themselves are personages, they will be more flexible, and perhaps more reflexive: they can see themselves as characters in a multi-authored story, rather than prime movers who mould the world and the word to their will. We would argue that actors will be more effective that way, or at least can then avoid being buried under the weight of circumstances and reactions that they had shovelled out of sight. We would like to argue that inchoate organizational realities can be addressed better through the second genre - realizing that this argument about how to be successful is itself phrased in a modernist vein. It is because of this conundrum, how to make a difference when one realizes that making a difference does not really depend on one's own action, that we discussed the relationship of text and action, of agency and narrative (see also Mike Michael's discussion on representation, perfomativity and materialisation in Chapter Two of this volume).

At a deeper level, agency of the actors is shown to be constructed through narrative. While agency as an independent source is an illusion, stories which introduce heroes and villains and thus create agency, and guide it along, have effect. In that sense, agency is a productive illusion. Some reflexivity is necessary to avoid becoming a prisoner of the illusion. The overall thrust and the narrative infrastructure is the outcome of such interacting narratives.

Contested Characters

A general reflexive lesson is the recognition of the duality of creating characters - THE USER, THE ALLY, THE PRODUCT-TO-BE, THE ADVERSARY - which are not only typifications but also actors/authors in their own right, which go their own way. While this can be read as simply saying that one cannot force others to do as one wants, the point is that actors often behave as if this were the case. The narrative shaping has a strong hold. It is through recognizing these mechanism, and in concrete situations, that the point is brought home. Meta-stories like the one we developed in this chapter contribute to this recognition, and stabilize it.

Thus, the recognition of the role of narratives in interaction is important, because it offers a handle on heterogeneity and ambiguity in the life of organizations in rapidly changing environments. Directly, in specific stories and interactions, because 'narrative permits ambiguity and enjoys paradoxes' (Czarniawska-Joerges, 1995, p.15). And over time captivity in an emerging path-dependence decreases when streamlined reconstructions of innovation journeys are recognized as effects of narrative infrastructure.

This conclusion, however persuasive and important, hinges on the existence of a boundary between the inside (where heterogeneity can be reduced with routines) and the outside, the external environment, full of strangers with their own visions. Thus, we need a second conclusion since narratives are not limited to one's own organisation but, instead, are implicated in the narratives of 'others'. On occasions, the narratives of such others will contest and destabilise an otherwise heroic production narrative.

In other words, the 'product triumphant' may be victorious on its own terms, but not necessarily in the wider world. This is a cautionary message to the enlightened modernist project managers. But the message works also in the other direction: the contested futures pressed by actors in public spaces produce an interesting spectacle, but is this more than a show for public audiences? In order to be effective, there must be links with product creation processes and with processes of embedding in society. Actors must realize that they are characters in the stories of future worlds put up by other actors/authors.

The notion of 'contested' futures then shifts from a battle of interests, with the scenarios, promises and risks as weapons in the struggle, to a recognition of narrative and narrative infrastructure as the environment (context, repertoire) through which actors define their preferred actions, and in which they position themselves and others. If this is the basic pattern, biotechnology firms (the small as well as the large variety), venture capitalists, retailers, consumer and environmental groups, all collude in creating a multi-actor - and multi-authored - story. Instead of becoming a victim of the tensions inherent in attributing praise or blame (as is common in controversies), one might go for re-description and conversation (Rorty, 1989) - provided one understands and accepts heterogeneity and the limited scope of a narrative in the context of wider narratives.

Acknowledgements

A previous version of this Chapter, entitled 'Narrative Infrastructure in Product Creation Processes,' appeared in *Organization*, 7, 1, 2000. The editors would like to thank Sage Publications for their permission to publish the revised version that appears here.

References

Boje, D.M. (1991) The Storytelling Organization: A Study of Story Performance in an Office-Supply Firm, *Administrative Science Quarterly*, 36, 106-126.

Callon, M. (1986a) The Sociology of an Actor-Network: The Case of the Electric Vehicle, in Callon, M., Law, J. and Rip, A. (eds) *Mapping the Dynamics of Science and Technology: Sociology of Science in the Real World*, Macmillan, Basingstoke/London, 19-34.

Callon, M. (1986b) Some Elements of a Sociology of Translation: Domestication of the Scallops and Fishermen of St. Brieuc Bay, in Law, J. (ed), *Power, Action, and Belief: A New Sociology of Knowledge?* Routledge, London, 196-223.

Callon, M. (1991) Techno-Economic Networks and Irreversibility, in Law J. (ed), *A Sociology of Monsters? Essays on Power, Technology and Domination*, Routledge, London, 132-161.

Callon, M (1992) The Dynamics of Techno-Economic Networks, in Coombs, R., Saviotti P. and Walsh V. (eds), *Technological Change and Company Strategies*, Academic Press, London, 72-102.

Callon, M., Law, J. and Rip A. (eds) (1986) *Mapping the Dynamics of Science and Technology: Sociology of Science in the Real World*, Macmillan, Basingstoke/London.

Czarniawska-Joerges, B. (1995) Narration or Science? Collapsing the Division in Organization Studies, *Organization*, 2, 11-33.

Deuten, J.J. (1994) *Ironie en Produktontwikkeling [Irony and Product Development]*, University of Twente, Enschede (Final Thesis for the Degree in Philosophy of Science, Technology and Society).

Deuten, J.J., Rip, A. and Jelsma, J. (1997) Societal Embedding and Product Creation Management, *Technology Analysis and Strategic Management*, 9, 219-236.

Greimas, A.J. (1987) *On Meaning. Selected Writings in Semiotic Theory*, University of Minnesota Press, Minneapolis.

Hughes, T.P. (1983) *Networks of Power: Electrification in Western Society, 1880-1930*, Johns Hopkins University Press, Baltimore.

Jelsma, J. and Rip, A. with the cooperation of Van Os, J.L. (1995) *Biotechnologie in Bedrijf: Een bijdrage van Constructief Technology Assessment aan Biotechnologisch*

Innoveren [Biotechnology in Business. A Contribution from Constructive Technology Assessment to Innovating in Biotechnoly], Rathenau Instituut, Den Haag.

Latour, B. (1984) *Les Microbes. Guerre et Paix. suivi de Irréductions*, A.M. Métailié, Paris.

Latour, B. (1992) *Aramis, Ou l'Amour des Techniques*, Editions La Découverte, Paris.

Law, J. (1994) *Organizing Modernity*, Blackwell, Oxford.

Rip, A. and Talma, S. (1998) Antagonistic Patterns and New Technologies, in Disco, C. and Meulen, B.J.R. (eds), *Getting New Technologies Together: Studies in Making Sociotechnical Order*, Walter de Gruyter, Berlin, 299-322.

Rorty, R. (1989) *Contingency, Irony, and Solidarity*, Cambridge University Press, Cambridge.

Van de Ven, A.H., Angle, H.L. and Poole, M.S. (eds) (1989) *Research on the Management of Innovation: The Minnesota Studies*, Harper & Row, New York.

Van Lente, H. (1992) De Dialectiek van Beloftevol Onderzoek [The Dialectics of Promising Research], *Kennis en Methode*, 16, 150-171.

Van Lente, H. (1993) *Promising Technology: The Dynamics of Expectations in Technological Developments*, University of Twente, Enschede (Ph.D. Thesis).

Van Lente, H. and Rip, A. (1998) Expectations in Technological Developments: An Example of Prospective Structures to Be Filled in By Agency, in Disco, C. and Van der Meulen, B.J.R. (eds), *Constructing Socio-Technical Order*, Walter de Gruyter, Berlin, 203-229.

Wagenaar, H. (1997) Beleid Als Fictie: Over de Rol van Verhalen in de Bestuurlijke Praktijk [Policy as Fiction: On the Role of Stories in Administrative Practice], *Beleid and Maatschappij*, vol. 24, 7-20.

5 Organising/Disorganising the Breakthrough Motif: Dolly the Cloned Ewe Meets Astrid the Hybrid Pig

NIK BROWN

The 'breakthrough' motif today serves as one of the most pervasive temporal abstractions for describing key events in science and medicine. Of late, it has come to refer to commemorative moments including the introduction of penicillin, antibiotics, x-rays, vaccination, radiation therapy, transplantation, new genetics and much more. And every breakthrough has its towering 'heroes' and 'pioneers': Pasteur, Fleming, Florey, Jenner, Curie, Crick and Watson, etc. As such, breakthroughs signify unequivocal disjunctures in the overall temporal shape of innovation and science, designating all the major 'steps forward' of grand progress narratives (Lyotard, 1984). It is therefore, probably one of the most routine cultural methods available for making tacit sense of the dynamics of change and the relevance of 'the new' to the future.

This is not to imply that breakthrough is unambiguously favourable to those who use it most. Both scientific institutions and popular science writers express competing points of view. The metaphor is sometimes derided by both constituencies as an overused cliché that inflates hopes and creates promises which too often go unfulfilled (Palevitz and Lewis, 1998; MacNair, 1995). A good example of this is a recent meeting between senior scientists and correspondents at Cold Spring Harbor entitled *'Breakthrough! How News Influences Health Perception and Behaviour'*. Illustrating the ambivalences embedded in breakthrough, during the course of the meeting, the metaphor was quickly abbreviated to the pejorative 'B-word'. In a review of that meeting by two of its attendees, both scientists, these tensions were expressed as follows:

87

The Use and Abuse of the 'B' word
The 'B' word - breakthrough - divides scientists and journalists as no other... no word better signifies the crosscurrents and undertows that can sink the communication process. And none better reveals the cultural divides that separate the two professions. Is the B word abused, to the extent that its impact is diluted? (Palevitz and Lewis, *The Scientist*, 20.7.98)

Yet, the metaphor is also valued for the very reason it is derided. It is held to be a convincing vehicle for disseminating findings, generating future patronage and legitimating funding (Kent, 1997). In another skirmish, a *British Medical Journal* author publicly criticised a press release she had received announcing 'a breakthrough for sufferers of Noonone Syndrome', taking issue with the press release's 'loss of perspective as to the importance of the discovery of the gene' (MacNair, 1995). The Director of a prominent patient advocacy organisation, the Genetic Interest Group (GIG) protested at this, writing:

I hope that Macnair's views regarding the press release announcing a 'major breakthrough for Noonan Syndrome' are not representative of those held at the *BMJ*. The discovery of a gene... is a major breakthrough for those at risk... It shows that progress is being made and provides hope for the future... If the release had been headed 'Minor advance for those with obscure disease - not many interested' I doubt that many people would have been moved to read it. The media make the rules... the *BMJ* should not take the high moral ground if those with something to say play by those rules (Kent, *BMJ*, 1995, 310, p.672).

Now neither of these interpretations contest breakthrough as such but merely comment on its proper application. In other words, the motif is used as an ideal measure or benchmark with which to judge events as being either hyperbolic or having a real 'future'. Apparently then, it is valued highly enough to merit protection from being sullied by misapplication. As the last extract clearly illustrates, breakthrough is also evidently the axis in a distribution of blame between two reporting constituencies, science and the media. For different groups, it therefore defines the limits and boundaries of what counts as good and bad discourse about knowledge, science and the future.

This chapter is concerned with doing something quite different to determining when events properly deserve to be graced with one of contemporary science's most cherished temporal abstractions. Instead, it sets out to open up 'breakthrough' to explore some of the rhetorical, historical and material constitutive properties of a metaphor that will always misrepresent the messy and indeterminate way in which knowledge is actually made and what it is capable of doing for the future. So unlike those constituencies described above, I am not concerned with redeeming some core breakthrough value to be preserved and held up as an ideal to which all disclosures in science and technology should aspire.

The chapter also shows how breakthrough has emerged alongside wider changes in the proprietorial, public and utility focussed character of science and, as such is intimately tied into the two-fold practical orchestration of present problems and future solutions. It identifies breakthrough as probably the most powerfully future oriented metaphor within the current disclosure repertoire of science and science journalism. In other words, it lends itself to the construction of a future in a way that other forms of disclosure representation, particularly the 'discovery' motif, do not. The importance of making sociological sense of breakthrough arises from its pervasive use in cultural discourse about science and technology and its hitherto absence as an object of analytical attention, at least in Science and Technology Studies (STS). Another compelling reason for this analysis is that despite its ubiquity, breakthrough is, as we will see, a very recent addition to scientific reportage. Addressing the metaphor in greater depth is all the more pressing since, as a number of studies have pointed out, public interest and confidence in science is seen to be most concentrated in those areas commonly considered 'breakthrough medicine' (Durant, Bauer and Gaskell, 1998).

Empirically, my argument is comparatively situated in two disclosure cases, Dolly the cloned ewe and Astrid the hybrid pig. When the Roslin Institute announced that they had produced a mammalian clone, they had no anticipation of the furore that would follow in Dolly's wake. The central problem was that Dolly came to represent a whole different universe of futures to the one that Roslin's researchers had in mind for her. As everyone knows too well, the controversy lies in the seemingly sudden capacity to create endlessly repetitious duplicates of otherwise distinctive beings. More recently though, Roslin's disclosure has also been contested in respect to whether and on what basis Dolly can legitimately considered to be a clone. In the other case, that of a hybrid pig, science's politics have been involved in different kinds of dispute arising from bodies that traverse species boundaries. Such hybrids cut through species difference with an unprecedented traffic in genes, tissues, cells, organs and even viruses. Like clones, hybrids too occasionally rise to the dizzy heights of meriting a name. Astrid was the first female transgenic pig produced by a British firm for organ provision in human replacement surgery. The case will illustrate the work done to qualify one of the firm's disclosures as a breakthrough in the immunological similarity of 'donor' animals and 'host' humans.

Fig 5.1 Breakthrough of the year 1997. From the cover of *Science* 19.12.97. Reprinted with permission. Copyright (1997) American Association for the Advancement of Science, and the artist, Ann Elliot Cutter.

The central problem for this analysis lies in how to make sense of the role of the breakthrough motif within contrasting contexts such as the two cases just introduced, asking: what and whose purposes it serves, what networks of activity are involved, on what basis breakthrough comes to be contested. Before exploring the Dolly and Astrid cases in more detail, there are a number of analytical approaches outlined below that will be of value in answering these questions. First, while there are no direct accounts of breakthrough in STS scholarship, there is a literature on breakthrough's near pseudonym 'discovery'. Contrasting the two terms is essential to determining what it means to invoke relatively distinct disclosure rhetorics. Second, breakthrough is inextricably wedded to the conventions of news discourse, the same conventions so frequently cited by scientists as being responsible for the ubiquity of the motif in the public communication of science. Finally, central to the construction of a breakthrough from past problems to future promise is the issue of timing. Here, I borrow on the rhetorical analytical term 'kairos', literally meaning 'the right time', asking how it is that timing contributes to the mobilisation of disclosure metaphors and futures.

Interpreting Breakthrough

Deconstructing Discovery Accounts

Woolgar and Brannigan, in separate projects, have both offered critiques of discovery episodes in which the metaphor is presented as an unstable practical-rhetorical achievement. Upon closer examination, discoveries are unstable both in respect to what really happened (Woolgar, 1976) and also the values or measures used to assess the significance of events in science (Brannigan, 1981). Woolgar's version of the 'discovery' of pulsars starts with inconsistencies between retrospective accounts by different members of the original research team based at the radiotelescopic observatory in Cambridge. These variations suggest that discovery is just one of many possible interpretations rather than being a singularly consensual description of 'what really happened' (Woolgar, 1976 p.395).[1] For Woolgar, 'discovery' implies an unrealistic commitment to 'preconceived notions of instantaneous discovery' rather than extended process, and this reduction in complexity increases as events recede further into the past. (ibid. p.417). Similarly, Brannigan critiques what he calls 'folk theories' of discovery where the idiom is attributed to, for example, inspired genius or cultural determination (Brannigan, 1981). The second of these explains discovery as an inevitable outcome of a culture's level of development producing the same results in scientists working independently at the same historical time (ibid. p.46). Other 'folk' criteria for discovery qualification include *originality* or the perceived precedence of an event; its validity in context, meaning that not all 'discoveries' were held to be such at the time original claims were made. Likewise, successful discoveries are sometimes disqualified when their premises are subsequently contested.

At one level, 'breakthrough' and 'discovery' might be said to be similar. Both unquestionably depend on a retrospective concentration within single events rather than process and both index time by recalling prior key moments with which to compare an event. Van Lente in Chapter Three of this volume makes a similar observation regarding the historically comparative properties that are constitutive of metatemporal discourses, particularly that of 'technological progress'.

Now, despite the slippage, breakthrough and discovery metaphors differ in some very important respects. The empirical cases discussed below indicate that in the case of breakthrough, this indexing of time is also prospective in that the metaphor implies the building of suspense and momentum towards future events and new impasses. Also, the discovery metaphor is more usually used to characterise the uncovering or *laying bare* of a universal nature (*new knowledge*) whilst breakthrough tends to be associated with novel solutions to existing well-defined problems (*new technologies*).

Probably one of the most important differences between the metaphors is that unlike discovery, breakthrough has only recently entered the lexicon of science disclosure. Taking my cue from William's keywords-type analysis, there's a striking etymological history in how breakthrough comes to be attached to science and technology in the way that it does today (Williams, 1983). At the turn of the twentieth century, breakthrough largely describes military campaigns, before being used to talk about the breaching of economic barriers and the creation of new markets. But it is not until as recently as the late 1950s that it first makes its appearance as a signifier for science and technology. In a telling blend of military and scientific reference, a feature in *The Listener* in 1958 describes 'The technological break-through which allowed both the United States and the U.S.S.R. to produce H-bombs within a year of each other' (11 Sept 376/2). So the common use of breakthrough to refer to events before this time, such as those listed at the beginning of this chapter, reveals a very recent historical revisionism of pre-mid twentieth century events in science and technology.

The significance of breakthrough's recency can not be overestimated since it registers far reaching changes in the way knowledge is represented, practised and perceived. Shifts in metaphors such as this are materially and institutionally very powerful (Lakoff and Johnson, 1980). For Fox Keller, the metaphor of nature's 'secrets' has long provided a motivation for science and the need for counter metaphors like discovery with which to 'probe' secrecy (1992). Discovery implies the chance giving up of nature's truths to the enlightenment's impartial observers, its 'modest witnesses'. In effect, this is nature doing what it normally does but now observed by scientific onlookers equipped with the experimental instruments to 'uncover' what was secret before: 'The ferreting out of nature's secrets, understood as the illumination of a female interior, or the tearing of Nature's veil, may be seen as expressing one of the most unembarrassedly stereotypic impulses of the scientific project' (ibid. p.41). The shift towards breakthrough from discovery, we might contend, signals a newly contemporary object for feminist critique. Breakthrough arguably represents a new and more aggressive repertoire. By necessity it implies the requirement of considerable force to push through a barrier of some kind: there is very little that is modest about that!

Also, the recent emergence of breakthrough certainly resonates with what Michael Gibbons and others have called the 'new mode of knowledge production' (1994). That is, actors are increasingly producing knowledge in the contexts of problem and application. To this extent, breakthrough has become the metaphorical location of values and activities whereby knowledge is rewarded and validated in relation to actual and clearly defined problems or impasses rather than, as in the case of discovery, being prized for its speculative or serendipitous character.

Deconstructing News

The deconstructing discovery story described above and embedded in the accounts of Brannigan and Woolgar tends to attribute temporal abstractions like discovery to disclosure by scientists and their institutions. But such abstractions, and breakthrough in particular, emerge in the interface with other forms of reporting too. As the disputes with which I began this chapter illustrate, it is the requirements of news discourse and the demands of the daily competition for the documentation of events that are widely seen (by many scientists at least) to be instrumental in shaping science disclosures into spicy breakthroughs. To count as news, suggest analysts like van Dijk and Bell, information should be novel, unprecedented and recent (van Dijk, 1988; Bell, 1995). The competitive dimension also imposes certain constraints in terms of length, brevity and the accessibility of news accounts to as wide a share of the viewing/reading market as possible. Like the breakthrough and discovery metaphors, news is temporally condensed or foreshortened in response to such pressures. News also relies on a certain degree of presupposition that interpretatively prepares the audience to receive information as news. Where preparation is seen to be inadequate the story will include a brief chronology spelling out how a crisis developed or a situation became more acute until the new/s event itself either exacerbates or resolves the developing narrative.

Not surprisingly, the breakthrough metaphor lends itself neatly to many if not all of these features. For example, interpretative preparation might mean a shared understanding that breakthrough properly describes how new knowledge comes about. Now, in contrast to the cited disputes in the *BMJ* and *The Scientist* with which I began this chapter, laying responsibility for breakthrough at the door of science correspondence falls far short of being an adequate explanation for how science disclosure comes to be breakthrough news. As the Dolly and Astrid cases will illustrate, the picture is far more complicated and ambiguous. Indeed, it is the ambiguity surrounding who is responsible for the social scripting of breakthrough that allows scientific and journalistic actors to exchange ideal reporting identities and conventions when it suits them to do so.

Rhetorical Analysis and 'Kairos'

The fate of a scientific disclosure rests upon the configuration of an appropriate temporal context in which its significance can be readily understood. 'Kairos' is a classical rhetorical term through which analysts have sought to understand the temporal context of a rhetor's interjection, how it is that people simultaneously construct and respond to a temporal context in which actions either succeed or fail. Kairos literally means *the right time... what makes this the right time?* It implies an occasion for agency that is specific not to any time, but to *this time* rather than another (Smith, 1969; Miller, 1992, 1994; Kinneavy, 1986). Very often the opportunity is a consequence of a problem having led to a crisis such that, for example, a particular scientific claim might be said to be 'timely' or even overdue. Miller has applied the term to Watson and Crick's 1953 disclosure in *Nature* of the molecular structure of DNA (Miller, 1992). She contrasts this with Oswald Avery's claim a decade earlier in which he similarly presented research evidence pointing to DNA as the biological agent in replication (Avery et al., 1944). Even though the conclusions were in effect almost identical, their timing was quite distinct with Avery's paper floundered in a knowledge community where there were few prepared to recognise the claim as the answer to a widely asked question. Kairos points to the temporally extended processes whereby expectations are circulated and come to converge upon a particular moment. Kairotic moments, practically and rhetorically organised, represent a concentration of agency, a disjuncture, in which it becomes easy to forget or hide the many other contingencies or agents upon which a 'right time' came to depend. Let me now consider some of these analytical resources through the two cases, beginning with a 'breakthrough' in the development of transgenic animals for human replacement surgery.

A Transgenic 'Breakthrough' - Astrid the Hybrid Pig

Astrid is one of a cohort of transgenic founder stock produced by the British biotechnology firm Imutran. Her genome contains the gene for a function of the human immunological system, Decay Accelerating Factor (DAF), the idea being that upon transplantation, the organs of her kind will fail to trigger at least one rejection process involved in human immunology. This case focuses less on Astrid per se and more on a disclosure event to which she was an essential precursor. In early September 1995, Imutran invited science correspondents to join them at a press conference at the Royal Society of Medicine on the 12th of that month. With details still undisclosed, Imutran's press office hinted at 'major new findings' and 'important progress'. In this way, a special moment was chosen to disclose experimental trials in which the hearts of ten cynomolgus monkeys were excised and replaced with ten transgenic pig hearts. At the event a document was distributed to attendees and forwarded to major news agencies around the world. This extract below is taken from the technical contents of that press release:

- Each received a transgenic pig heart and was given similar levels of immunosuppression as humans.
- Of the 10 transplants, 2 are currently surviving at up to >60 days.
- Examination of two monkeys on days 34 and 35 with the pig hearts still beating showed that the hearts were normal with no signs of rejection.
- The median survival for this group is currently >40 days.
- Control hearts survived 55 minutes.

The rest of the press release sets out to translate what this data means such that otherwise relatively obtuse technical information is put into interpretative context:

> The success of the trial... confirms that the technology could be the answer to the current organ donor shortage. Imutran believes its technology is now ready to be tested in humans and expects to begin the first trial in 1996, in the UK. Studies will be carried out at Papworth Hospital... .

Imutran has overcome the major hurdle in the development of animal organs for transplantation into humans... . This contrasts with work carried out by a group in the USA... using similar technology, which recorded a maximum survival of only 30 hours... .

Director of Cardiac Transplantation at Papworth Hospital added his endorsement for Imutran's *ground-breaking* work. 'This research is now well advanced and we are making excellent progress... The programme of human clinical trials planned for 1996 will be a big step forward... a genuine advance in transplantation' [my emphasis].

First, the document describes the current impasse to which these findings represent a breaching, that being an 'acceptable' level of immunological parity between a 'donor' and 'host' species. This is then used to signal the possible breaching of a still present impasse, the xenogeneic solution to the shortage of replacement tissues and organs for human replacement surgery, a future breakthrough. The press release repeatedly translates the animal trials into terms that anticipate the future clinical trials on human patients then scheduled for 1996. This 'breakthrough' is then infused with descriptions of 'genuine advance', 'big step[s] forward', 'excellent progress' and the distant prospect of the 'potential to save lives', all essential to creating the suspense around a future 'right time' a future 'breakthrough'.

In this way, Imutran's breakthrough depends on the experiment, and all that led up to it, extending over months if not years, being truncated and compressed into the terms of a momentous announcement at a press conference and through a press release. The moment of the disclosure itself is just as important. The press conference is scheduled to coincide with the annual meeting of the British Association for the Advancement of Science such that the meeting itself is used as an additional public platform for the breakthrough:

The announcement was described as highly significant by a leading transplant surgeon attending the British Association for the Advancement of Science annual conference... (*The Times*, 13.9.95).

In all, the announcement on September 12th translated into an evenly prominent media story. Whilst some of the coverage was scattered around the general date of the disclosure, most reports clustered on the day following, the 13th, in keeping with the recency ethos of news. In all, the immediate press response to Imutran's disclosure was invariably cast within the temporal terms of 'breakthrough'. A brief review of the content of the media coverage echoes the press release and is concentrated around several main themes. First, Imutran's breakthrough is routinely compared with other salient moments of therapeutic efficacy. In this way, analogies between historically separate events identify the current breakthrough as proportionate in significance to other breakthroughs:

> It is the most exciting breakthrough since the first heart transplant operation was performed by Christian Bernard in 1967 (*Daily Mail*, 13.9.95).

> The breakthrough is regarded as the biggest advance in transplants since the introduction of the drug that suppresses organ rejection 10 years ago (*The Telegraph*, 13.9.95).

The metaphor can be either explicit as is the case in these extracts or implicit by being evocative of all those historical referents with which this event can be associated. It also entails a moral imperative such that if this event is as significant as 'the first heart transplant' then any opposition to the current development is proportionate to having halted the historically distant technologies from which we are currently seen to benefit. In this way, the breakthrough is intended to foster a more conducive regulatory context for a technology which might otherwise be more vigorously contested.

Secondly, the representation of the Imutran breakthrough also assumes a naturalised or 'black boxed' rendering of the xenotransplantation solution route. This effectively endorses the indivisibility of the fate of the technology and the fate of patients at the mercy of 'the critical shortage in replacement tissues and organs'.

> This week Imutran... said it had successfully transplanted pig hearts into monkeys... . [Terence English:] we still seem to be some years away from a reliable, cheap, totally implantable mechanical device that will take over the action of the human heart. It is not surprising that in the last few years there has been intense interest in the possible application of 'xenotransplantation'... (*The Guardian*, 25.9.95).

> More than half of the 5,000 people waiting for transplants die every year because no human organs are available. Answer: Pigs are now seen by many doctors as the answer to the acute shortage of donors (*The Daily Mirror*, 24.9.95).

Of course, the shortage crisis does not unproblematically equal an XTP solution. Other technologies might just as easily compete for ownership of the 'organ crisis' including some cloning technologies, mechanical devices and policy changes like the principle of presumed consent to organ donation. But naturalising the XTP solution route has been the focus of considerable promotional endeavour illustrated by announcements like this.

Finally, the xenotransplantation breakthrough is represented as a significant, but nevertheless, preparatory moment for the future breaching of a current impasse:

> Transplant patients could be given hearts within a year following a breakthrough in genetic engineering.... Papworth surgeon John Wallwork, who is likely to perform the first operation, said: 'the programme of human clinical trials planned for 1996 will be a big step forward in the development of a genuine advance in transplantation' (*Today*, 13.9.95).

> Breakthrough could end transplant delays [headline]. Pigs' hearts could be given to humans early next year following a research breakthrough. 'If trials are successful we could end the lottery for life which at the moment means some patients remain sick, some receive organs and some die,' said... pioneer (*The Daily Express*, 13.9.95).

> Breakthrough enables trials to start next year [headline]. ...a consultant at the Freeman Hospital... said: 'The transplant community is waiting with bated breath for the case to be proved in clinical trials...' (*The Times*, 13.9.95).

The breakthrough, then, is not simply celebrated as a single, cumulative great achievement, but rather it is put to work in the articulation of a distant temporal horizon - the creation of suspense. It is represented as an ordinal position in time that remembers and constructs past and future breakthroughs. As it happens, the clinical trials set for 1996 were poorly timed to coincide with the height of the BSE/CJD crisis and general anxiety surrounding the risks of transpecies disease arising from this and similar technologies. To date, this has arrested the technology and shifted the locus of the impasse from immunological parity to overcoming viral pathogenic risk.

A Cloned 'Breakthrough' - Dolly

Dolly did not begin her public life until seven months after her birth when Ian Wilmut and fellow researchers published their article in *Nature* (Wilmut et al., 1997). As is the wont of scientific reportage, the paper was in all respects a dry technical retrospective account of the reproductive process leading to Dolly, cloned from the nucleus of a ewe's udder cell. It was not until much later that number 6LL3 would be renamed Dolly, a schoolboy reference to her somatic origins and the ample bosom of a US Country Western singer! Notwithstanding the joke, there are deep ambivalences in this substitution of signs. The giving of a name in the place of a number, as laboratory ethnographers note, makes relationships with research animals more tolerable, even socially satisfying (Lynch, 1988; Arluke, 1988).

Now Dolly was far from being any ordinary breakthrough. In a rather odd reframing of the beauty contest, she was actually voted 1997's 'Breakthrough of the Year' by *Science,* the journal of the American Association for the Advancement of Science (278, pp.2071-2192). The cover of the issue depicts Dolly standing in the foreground with a pyramid of her replicants receding into future time and space behind her. In effect, the temporal composition of the image represents Dolly glancing back over her shoulder to the future exponential duplication of her nuclear genome (see Fig 5.1). The honour itself was represented as something of a breakthrough by observers noting the rarity with which non-US-based achievements find their way into *Science's* 'top 10'.

> Sheep clone tops list of year's firsts in science [Headline]
> DOLLY the sheep has gone to number one in the list of top ten breakthroughs of the year, beating the Mars Pathfinder mission into second place. Heading the list as 'Science's 1997 Breakthrough of the Year' is Dolly, the world's first cloned adult mammal. The cloning of Dolly provoked the questions: do ethical concerns outweigh the possible social benefits of cloning? Can human cloning be far behind? (*The Daily Telegraph*, 19.12.97)

Whilst there is no mention of the breakthrough metaphor in the paper published in *Nature* by Wilmut et al., their engagement with wider publics and the press in particular is shot through with reference to the breakthrough significance of mammalian cloning by nuclear transfer. The following extract is taken from the press release sent out by the Roslin Institute and its commercial arm PPL Therapeutics. Note the standard use of the embargo preventing any accredited recipient pre-empting the carefully timed simultaneity of the disclosure before exactly 19.00hrs on 26.2.97. The use of the embargo is a routine way of making sure that the press has time to prepare copy whilst also coordinating the exact timing of the disclosure to coincide with some other event. In this case, the embargo anticipates, by no more than twelve hours, the issue of *Nature* in which Wilmut and colleagues publish their findings:

 News Release

**EMBARGOED UNTIL 19.00 HRS GMT, 26 February 1997
released 24 February 1997**

**Scientists at the Roslin Institute Publish
Scientific Breakthrough**

**ability to clone sheep through nuclear
transfer from somatic cells**

However, Dolly presents a number of problems to the breakthrough ideal. Like Astrid, she is embedded in temporal processes extending heterogeneously through many experimental episodes and across many different animal bodies. She was in fact the 277th attempt at producing a clone by nuclear transfer. Her timed introduction into the spectacle of public life on a specific date and at a particular chosen moment in the duration of Roslin's experimental activities abbreviates many of the contingencies and modalities that might otherwise disturb her breakthrough status. We might think of this as a truncation of time that squeezes experimental process into the breakthrough abstraction. And yet very little of this concentrated timing, this removal from modality, has been enough to protect the breakthrough from closer scrutiny, upon which we find that Dolly is indeed far from being the clone we all took her to be.

Dolly embodies the genetic attributes of multiple originals and is *not* exclusively a copy of one animal alone. To understand how, we have to go back the claims made in the original disclosure. In their paper, the Roslin Institute set out to explain how they had replicated a 6-year-old adult ewe by means of nuclear transplantation. That is, the nucleus of a somatic udder cell was taken from one ewe, treated to 'forget' its specific cell function and then transferred into an enucleated egg (oocyte) from a second ewe. The developing animal was then gestated in vivo in a third ewe before being brought to full term in fourth, Dolly's second surrogate parent.

Whilst it was relatively widely accepted that Dolly had inheriting the genetic attributes of the nuclear donor, it quickly transpired that she had also inherited the mitochondrial genome of the egg donor. The mitochondrion of a cell is responsible for making cells function properly and is located outside the nucleus. As such it has a separate genome to the nucleus and for this reason it is inherited matrilineally during fertilisation whether by heteroreproduction or nuclear transfer in this case (Evans, 1999). So Dolly inherits the mitochondial DNA of the egg donor and not the nuclear donor. In addition, she has also inherited genetic attributes and immunological properties from her two different *in vivo* surrogates.

As with the XTP case, the whole process of maintaining Dolly's breakthrough is ongoing. Probably one of the most striking illustrations of her ambiguous status revolves around claims that her genome is expressing internally different rates of aging. The proteins that protect the ends of her chromosomes (tolemeres) are shorter, in effect older, than they should be. The theory is that these structures continue to shorten throughout an organism's life eventually instructing cells to die. The length of Dolly's tolemeres seem to be inconsistent with her own age but consistent with that of the nuclear donor suggesting that she is simultaneously four and nine years old (*Nature*, 27th May 1999). This destabilises the whole discourse of biological 'copying' because Dolly does not *take after* the nuclear donor but in effect *is literally* the nuclear donor and thus subject to the same processes of aging and decay.

All of these factors have tended to undermine this event's precedence by extending the processual contingencies embodied in Dolly and reflected in the innovation history of the Roslin Institute and PPL Therapeutics. In another example, in the year preceding Dolly, the Roslin Inst. announced that it had produced two other cloned ewes, Megan and Morag, this time by dividing embryo cells ('blastomere separation'). Here, the (undifferentiated) cells (blastomeres) of a fertilised egg are separated to continue developing normally and separately. In fact, this form of cloning was used as early as 1992 in experimental human assisted conception but without allowing the embryos to continue developing (Hall et al., 1993). The research team at Roslin were as bewildered at the absence of excitement surrounding Megan and Morag as they were at its intensity surrounding Dolly. What distinguishes Dolly from the techniques of embryo splitting lies in intention. Split embryos are not intentional replicants of an already known animal because they still combine the gametes of two parents. Although, in fact, even the possibility of 'deliberate intention' must surely be questioned given all the grounds upon which Dolly is now suspected of not meshing with what the troubled word 'clone' is taken to mean.

These ambiguities around whether Dolly is conventional or novel and the connection with known intention are rife in what became the Dolly debate. It leads to whole different terms of reference. In avoiding the potentially pejorative term 'clone', the National Bioethics Committee of the US, refers to Dolly as the 'delayed genetic twin of an adult sheep'. Further illustrating the importance of intention, another analyst observes that 'Dolly was a deliberate copy of an adult animal, brought into being after her genome source had fully developed as an adult, this seems an inappropriate use of the term twin. Replicand seems more accurate...' (Baird, 1999 p.181).

Essentially, what these disputes illustrate is the perniciousness of the breakthrough as its modalities are opened up for detailed examination by participating constituencies. But as we all know, the instability of the breakthrough status was nothing by comparison to the controversy generated as the disclosure left the technical pages of Wilmut's paper in *Nature* and entered wider public debate where it quickly translated into a rehearsal of monstrous human reproductive futures.

Probably most significant here was the marked absence on the part of the Roslin team of any expectation that their research would lead to these kinds of questions. Rather, in the narrow framing of Roslin's innovation agenda, Dolly's meant nothing more than an empirical application of the theory of replication by nuclear transfer for the purpose of producing research and farm animals in a more efficient and exacting manner.

Much to the surprise of Roslin's researchers, this breakthrough 'science of similarity' has been widely taken to breach a political and cultural commitment to individual difference (a 'politics of difference'). The debate was, and is, infused with sanctions against improper substitution. In an dazzling blend of individualisation and geneticisation, the taboo prescribes that 'individuals' must be valued in their own right and that they should not bear the value (specifically in this case the genetic value) of another. Of course, the rest is history, albeit a continuing and fractious history.[2]

Broken Breakthroughs: Agency, Attribution and Blame

Cases like these raise a number of interpretative tensions in the analytical repertoire we use to interpret breakthrough in modern science and technology. Key questions centre on how we make sense of its textual and practical orchestration or contestation and how we conceive of the changing relationships between different reporting constituencies. Evidently, this is largely a problem of the attribution of agency, begging the question whose agency is reflected in the way the breakthrough metaphor comes to be attached to a particular event or technology? In other words, how is it that breakthrough comes to act a little like an agent itself by mediating the force and inertia of an innovation agenda (see also Deuten and Rip in Chapter Four)? Another property of these cases highlights the usefulness of interpretative ambiguity which allows reporting constituencies to apportion responsibility to one another and shift identities when, as often happens, disclosures fail to match the ideals embedded in breakthrough.

Breakthrough as Scientific Representation?

The interpretative ambiguity of agency and attribution is inherent in the analytical repertoire of STS itself. The deconstructing discovery literature of Brannigan and Woolgar leads us towards the representations of scientists as the authors of breakthrough. For example, Imutran enrol the press to address a much wider 'public' audience than they would otherwise have had access to. They achieve their breakthrough by compressing the extended vagaries of laboratory practice into the spectacular performance of the press conference. The Dolly breakthrough too is a consequence of the Roslin Institute presenting a truncated narrative of experimental process reported in *Nature*.

Both cases clearly illustrate the temporal reporting requirements of science. Take for example, the use of delay and deferral separating experimental events from journal publications and embargoed press releases. This is a characteristic property of scientific representation and an essential temporal tool in the production of certainty and therefore prestige and patronage by removing the situated modalities and process of knowledge production (Nelkin, 1995, 1996). Scientific news is therefore invariably old news made new for the purposes of reducing contingency and creating suspense toward some future horizon of action.

This discussion also highlights breakthrough as a distinctive motif in the broader repertoire of science disclosure and communication, particularly in contrast to discovery. Breakthrough is shot through with a problematised future in the way the discovery is not. It presupposes what Gibbons has called the problem context, or rather, the metaphorical shift towards a need to represent science as an instrumental knowledge activity solving applied problems (1994). Discovery on the other hand is linked more closely to serendipitous 'blue skies' ideals of what scientific knowledge creation should look like.

The centre of the axis between the motifs is the difference between *biology* and *biowealth*, that being the discursive requirements of 'knowledge for itself' discourse as opposed to proprietary discourse respectively. In his study of two research scientists, Myers observes their adaptation from the language of discovery in writing journal articles to learning the skills and language of invention in patent claims (Myers, 1995). Discovery is, loosely speaking, the province of the scientific article, a recognition of something already in nature whereby 'the scientist' is rewarded with a prize, continued funding or a fellowship in the Royal Society. Breakthrough is, again loosely speaking, the province of invention, application, utility, non-obviousness, the creation of a novel thing whereby 'the inventor' is rewarded with a patent, commercial sponsorship and perhaps an appointment to the executive board of a small British biotechnology company called either Imutran or PPL Therapeutics. In the former context, there is a problematised present or a natural anomaly that betrays the existing theoretical models of how something should in nature behave but does not. In the latter context, there is a problematised future, an impasse which the application's innovators promise to breach through inventive skill not found, *indeed must not be found*, in nature. Dolly and Astrid are not presented as something 'already out there', if that were so then it would undermine the claims to invention embedded in the discursive requirements of *biowealth* as opposed to *biology*. Neither of these identities or discourses is necessarily mutually exclusive, but can be deployed strategically and usually simultaneously to satisfy the requirements of different audiences.

Both cases neatly illustrate the requirement for a clearly defined connection between the innovation solution and widely accepted problems. The difficulty in the Dolly event was that her solution value was far from immediately obvious. She was not the answer to wider social questions about how to resolve a specific and agreed upon problem. Instead, the Roslin Institute were responding to questions confined within the expert scientific community about the hitherto theoretical but unproven feasibility of replication by nuclear transfer from an adult somatic cell. In the absence of 'a problem' her future was ambiguous and in the event swung easily from a fixed utility agenda to the potential application of nuclear transplantation to human reproduction. It was after the firestorm of debate had already begun that the Roslin Institute began to grasp the need for Dolly's practical value to be widely disseminated. In the case of Astrid, the connection between problem and the value of the experimental trials as a solution could not have been more obvious and went largely uncontested even if, at a later date, the promise became unsettled. So the mobilisation of breakthrough depends upon a successful problem definition that applies equally between the expectations of an expert or technical constituency and widely held cultural understandings of the utility value of risky innovations. Where there are significant differences between scientific and cultural definitions of the relationship between problems and solutions in the modern biosciences, acute moral questions are sure to arise.

Breakthrough as Cultural Representation?

On the other hand, the Imutran and Roslin disclosures are enrolled into the purposes of the press. Clearly, we need to supplement scientific representation with the temporal requirements of news discourse. Evidently, breakthroughs do not necessarily imply a relationship of mutual collusion between science and the press. There are important areas of conflict. Scientific communities blame news broadcasters for inappropriately presenting findings as breakthrough. Then again, science correspondents often complain of not having sufficient resources to screen press releases to present more cautious readings when promises subsequently go unfulfilled. These tensions were particularly evident in the Dolly case illustrating the complete inadequacy of Roslin's scientific breakthrough reporting to inoculate itself from its very own cultural politics. The technical character of the disclosure in *Nature* entirely failed to contain or police the technique's application to human rather than nonhuman futures.

Whilst certainly appealing to scientific groups, breakthrough often interferes with the ability of research communities to revoke findings later without losing trust (*Nature*, 393, 97, 1998). Revocability is built into the reproduction of the need for new knowledge but is clearly at odds with the definitive requirements of news discourse and the putative appetite for consistency both amongst political actors and publics at large (Yearley, 1989, 1995; Nelkin, 1995).

Now like the repertoires of science as invention and science as discovery, neither of these forms of representation belongs either to the media or science alone. Instead, they serve as ideal reporting values to which actors can lay claim to fulfil specific and situated disclosure needs. Scientific institutions and science correspondents routinely evoke the breakthrough motif when seeking to attract the interest of wider audiences. In so doing both lend credence to a culture which they may subsequently criticise when claims are revoked or judged to be hype. So, while clinging precariously to the idea of separate reporting values, the ambiguities of authorship allows different constituencies to blame each other when breakthroughs renege on promises or represent ethically difficult futures.

Breakthrough Subverted - From Textualisation to Socio-Materiality

Yet the discourse of breakthrough is implicated in other kinds of ordering besides those of the two reporting constituencies just discussed. We have to consider what kinds of socio-material timings or temporalities are being produced and how commensurate or incommensurate they are with one another. Breakthrough is a specific sort of time, or rather it is a product of a particular kind of sorting. To borrow Bruno Latour's maxim 'it is the sorting that makes the time, not the time that makes the sorting' (Latour, 1993 p.76). So other sortings can just as easily unmake breakthroughs or translate them into unintended or unforeseeable outcomes.

'Kairos', whilst emphasising the rhetor in the construction of a 'right time', also points to the broader socio-material heterogeneity in which breakthroughs are orchestrated. So it is far from adequate to account for the fate of breakthroughs in the terms of the textualised reporting conventions of science and the media alone. The foresworn clinical trials of Imutran clearly clashed head on with the timing of other kinds of sorting, particularly those of transpecies pathogens including CJD, BSE and even speculation on the origins of HIV. Breakthroughs are therefore rarely protected by the truncation of process and the removal from contingency. Dolly's inheritance is heterogeneous (or 'heterogenus' rather) and not 'monogenus'. Her production is uncertain and leads in many contrary directions. Astrid genetically embodies only one signifier of human immunity and there are many more both known and unknown involved in the rejection of tissues and organs, not to mention relative rates of aging between source and host species.

So in all, both science institutions and science correspondents are often responsible for presenting knowledge in the form of a metaphor that misrepresents the extended processes and contingencies involved in the production and value of experimental findings. Understanding the temporal dynamics around breakthroughs points to extended process over time and operating to different temporal principles rather than singularly momentous commemorative histories. It also demonstrates the tensions and opportunities present in the temporal terms of reference of different reporting constituencies in materially heterogeneous contexts of process. Finally, in keeping with the fine tradition of word-playing on pigs and sheep in contemporary biopolitics, I might plausibly be forgiven for concluding that:

> *Dolly udderly isn't a breakthrough clone and*
> *pig's won't fly even if ewe say they can!*

Acknowledgements

I would like to thank participants in two conferences where previous versions of this discussion were presented: *On Time, History, Science and Commemoration,*

The British Society for the History of Science annual conference, University of Liverpool, 16-18 Sept 1999; *Making Time/Marking Time*, British Sociological Association annual conference, University of York, 17-20 April 2000. Fig 5.1. is reproduced with permission from AAAS and the artist, Ann Elliot Cutting. Thanks also to fellow editors of *Contested Futures* and colleagues in the Department of Sociology at the University of York.

Notes

1. Deuten and Rip, writing in Chapter Four of this volume, also address the way reductions in complexity and contingency are superimposed onto historically distant objects (a GMO product in their case) through retrospective story telling.
2. For an excellent discussion of the way the Dolly debate has been conducted in the press and popular scientific commentary, see Franklin, 1999.

References

Arluke, A. (1988) Sacrificial Symbolism in Animal Experimentation: Object or Pet? *Anthrozoos*, 2, 97-116.

Avery, O.T., Macleod, C.M. and McCarty, M. (1944) Studies on the chemical nature of the substance inducing transformation of pneumococcal types. *Journal of Experimental Medicine*, 79, 137-57.

Baird, P.A. (1999) Cloning of Animals and Humans: What should the Policy Response Be? *Perspectives in Biology and Medicine*, 42, 2, 179-193.

Bell, A. (1995) News Time. *Time and Society*, 4, 3, 305-328.

Brannigan, A. (1981) *The Social Basis of Scientific Discoveries*, Cambridge University Press, Cambridge.

Dijk, T.A. van (1988) *News as Discourse*, Lawrence and Erlbann Associates, Holland.

Durant, J., Bauer, M.W. and Gaskell, G. (1989) *Biotechnology in the Public Sphere. A European Sourcebook*, The Science Museum, London.

Evans, M.J., Gurer, C., Loike, J.D., Wilmut, I., Schnieke, A.E., and Schon E.A., (1999) *Nature Genetics*, 23, 1, 90-93.

Franklin, S. (1999) What we know and what we don't about cloning and society, *New Genetics and Society*, 18, 111-121.

Gibbons, M (ed) (1994) *New Production of Knowledge: Dynamics of Science and Research in Contemporary Societies*, Sage, London.

Hall, J.L., Engel, D., Gindoff, P.R., et al. (1993) *Experimental cloning of human polyploid embryos using an artificial zona pellucida*. The American Fertility Society, co-jointly with the Canadian Fertility and Andrology Society, Programme Supplement, [Abstract of the Scientific Oral and Poster Sessions, 0-001, SI].

Keller, E.F. (1992) *Secrets of Life Secrets of Death*, Routledge, New York.

Kent, A. (1997) Letters: press release of the week. *The British Medical Journal*, 5, 310, 672.

Kinneavy, J.L. (1986) Kairos: A neglected Concept in Classical Rhetoric. In Jean Dietz Moss (ed), *Rhetoric and Praxis: The Contribution of Classical Rhetoric to Practical reasoning*, Catholic University of America, Press Washington DC, 79-105.

Lakoff, G. and Johnson, M. (1980) *Metaphors We Live By*, University of Chicago Press, Chicago.

Latour, B. (1993) *We Have Never Been Modern* (trans. C. Porter), Harvester, Wheatsheaf, London.

Lynch, M. (1988) Sacrifice and the transformation of the Animal Body into a Scientific Object: Laboratory Culture and Ritual Practice in the Neurosciences, *Social Studies of Science*,18, 265-89.

Lyotard, J-F. (1984) The Postmodern Condition (1979) (trans. G. Bennington and B. Massumi), Manchester University Press, Manchester.

MacNair, P. (1995) Medicine and the media. Press release of the week. *The British Medical Journal*, 5, 210, 67.

Miller, C.R. (1992) Kairos in the Rhetoric of Science. In Witte, S.P., Nakadate, N., and Cherry R.D., (eds). *A Rhetoric of Doing: Essays on Written Discourse in Honour of James L. Kinneavy*, Southern Illinois UP, Carbondale.

Miller, C.R. (1994) Opportunity, Opportunism and Progress. Kairos and the Rhetoric of Technology, *Argumentation*, 8, 1, 81-96.

Myers, G. (1995) From Discovery to Invention: The Writing and Rewriting of Two Patents. *Social Studies of Science*, 25, 57-105.

Nelkin, D. (1995) *Selling Science: how the press covers science and technology*, 2nd ed. W. H. Freeman, New York.

Nelkin, D. (1996) Medicine and the Media: An uneasy relationship, *The Lancet*, 347. 1600-03.

Palevitz, B.A. and Lewis, R. (1998) The use and abuse of the 'B' word, *The Scientist*, 12, 15.

Smith, J.E. (1969) Time, Times and the 'Right Time': Chronos and Kairos, *The Monist*, 53, 1-13.

Williams, R. (1983) *Keywords*. Oxford, University Press Oxford.

Wilmut, I., Schnieke, A.E., McWhir, J. et al. (1997) Viable offspring derived from foetal and adult mammalian cells, *Nature*, 385, 810-13.

Woolgar, S. (1976) Writing an Intellectual History of Scientific Development: The Use of Discovery Accounts, *Social Studies of Science*, 6, 395-422.

Yearley, S (1989) Bog standards: science and conservation at a public enquiry. *Social Studies of Science*, 19, 421-438.

Yearley, S (1995) The Environmental Challenge to Science Studies. In Jasanoff, S., Markle, G., Person, J. and Pinch, T. (eds) *Handbook of Science and Technology Studies*, 457-79.

6 Talking About the Future: Metaphors of the Internet

SALLY WYATT

Thinking About the Past: Metaphors and Electricity

'The whole universe is probably one almighty power station,' said Peter Fisher - more patiently, for my sake. 'We speak of the current, or flow, of electricity, but that is a metaphor.'

'Well then,' said my father, 'if you electricians take as much of this metaphorical stuff out of the air as it seems you intend to, you will upset the balance of nature. It stands to reason. You are endangering life upon earth. Everything will run down. It will mean the end of civilisation, a return to barbarism.'

'Electricity is not consumed like gas or oil. It does its work and then - well, I suppose you could say it goes on its way, passes on.'

Mother was looking more than usually puzzled. I knew that this was because she was not sure what a metaphor was. I told her in a quiet voice that a metaphor was describing something in terms of something else.

'Why should anyone want to do that?' she asked me. ...

'Either because you cannot find any other way to express what you mean, or in order to make a poetic or colourful effect. Like when you say God is love, or the cats are the very devil.'

Mother turned to Peter Fisher:

'In what way is the flow of electricity a metaphor? What is it really, if not a fluid?'

He breathed deeply. 'Electricity,' he said, 'is a medium of communication between two objects.'

'Or two people?' I asked.

He looked at me. 'In certain circumstances.'
(Glendinning, 1996 pp.15-16).

In this chapter, I examine some of the metaphors that are presently being used to describe the Internet in order to understand the perceptions and expectations of some of the actors involved in shaping the Internet. The

Internet is not yet a stable technology, in the broadest sense of the term 'technology': its technical features are changing (though not necessarily all of them, nor at the same rate) and its uses are highly variable. There remains a great deal of interpretative flexibility regarding what it is, what problems it can solve and what problems it may create. The Internet is not simply bandwidth, routers and servers; it includes the social relations associated with the production and use of this network of networks (Thomas and Wyatt, 1999). Because of this instability and uncertainty, policy-makers, industry spokespeople, journalists and academic commentators often deploy metaphors in order to convey their image of what the Internet is or might be. The future has to be discussed in terms of the imaginary, in terms of metaphors. Sometimes, today's imaginary becomes tomorrow's lived reality. Therefore, it is important to think about metaphors of the Internet not only because they reveal what different actors think it is but also because they tell us something about what they want it to become. For example, those who use metaphors of consumption generally and shopping malls in particular will devote resources to developing secure exchange mechanisms. Broadcasting metaphors carry with them assumptions about the nature of interaction between audiences and content providers that are more passive than those suggested by interactive role game metaphors and applications.

The opening extract from Victoria Glendinning's historical novel, *Electricity*, forces us to reconsider our understanding of the relationship between metaphors on the one hand and science, technology and the real on the other. Glendinning's novel is set in England during the 1880s. Throughout the eighteenth century and the early part of the nineteenth century, electricity was conceptualised as a fluid and thus the language used for describing the movement of water was adopted to analyse the movements of electricity. By the middle of the nineteenth century, scientists were less certain of the nature of electricity. If liquids and gases were atomic in structure, then perhaps electricity was too. During the same period, there was also much activity about the transmission of sound and the nature of sound waves. Scientists were investigating cathode rays which carried electric charges, but were uncertain about whether these most closely resembled oscillating waves, like those identified by Hertz, or whether they were streams of particles. In 1897, J.J. Thomson settled the debate in favour of the latter: electrons flow. Peter Fisher, a major character in Glendinning's novel, is correct to point out the metaphorical nature of discussing electricity in terms of currents and flows. Electricity provides a good example of the recursive relationship between language and the reality it is attempting to describe. Metaphors not only help to make science comprehensible to non-scientists, they can also guide scientific work.

Metaphors are not the sole preserve of poets and writers of magical realism; nor is their use simply an innocent attempt by commentators or politicians to demonstrate their own imaginative capacities or to appeal to the imaginations of their audiences. The previous paragraph demonstrates that metaphors are not only descriptive; they may provide clues to the design intentions of those who use them and as such, they may help to shape the cognitive framework within which actors operate. Metaphors also have a normative dimension; they can be used to help the imaginary become real or true. Friedrich Nietzsche described truth as, '[a] movable host of metaphors, metonymies, and anthropomorphisms: in short, a sum of human relations which have been poetically and rhetorically intensified, transferred, and embellished, and which, after long usage, seem to a people to be fixed, canonical, and binding. Truths...are metaphors that have become worn out and have been drained of sensuous force' (in Breazeale, 1979 p.84). Different social groups use different metaphors to capture and promote their own interests and desires for the future. Highways, railroads, webs, tidal waves, matrices, libraries, shopping malls, village squares and town halls all appear in discussions of the Internet. 'Windows' and 'menus' have already been incorporated into the language of Microsoft users with their misleading suggestions of choice, transparency and openness. Not all metaphors are equal. George Lakoff and Mark Johnson go further than Nietzsche and observe that, '[n]ew metaphors...can have the power to define reality. ... [W]hether in national politics or everyday interaction, people in power get to impose their metaphors' (1980 p.157). Thus, there are both cognitive and normative dimensions of metaphors which need to be considered: metaphors may convey something about the future functions and technological configurations of the Internet and they may also reveal the political assumptions and aspirations of those who deploy them. It is important to remember that it is social actors who exercise agency in their choice of metaphor. However, as metaphors become embedded within discourses and as actors become less reflexive in their use and choice of metaphors, it may appear that the metaphors themselves are becoming active agents carrying with them expectations about the future. Thus, it may be more appropriate to consider metaphors as discursive elements mediating between structure and agency.

We need to be careful with metaphors. They can help us to comprehend the new, the unseen, the unknown; but they can also mislead - sometimes deliberately - because the kinds of experience they purport to connect may be incommensurate. Terry Eagleton makes this point eloquently, '[b]oth history and nature are matters of process, to be sure; but to over-emphasise this is to risk eliding the distinctions between them in positivist or idealist style. A river does not flow as a sonnet does, nor does time fly like a goose' (Eagleton, 1997 p.22). The danger of elision is not sufficient reason to eschew either creating or analysing the metaphors at work in our world. Instead it means we need to recall D. McCloskey's advice concerning metaphors in social science: 'Self-consciousness about metaphor would be an improvement on many counts. Most obviously, unexamined metaphor is a substitute for thinking - which is a recommendation to examine the metaphors, not to attempt the impossible by banishing them' (1986 p. 81, cited in Joerges, 1989 p.48).

In the remainder of this chapter, I shall first return to the description of metaphors provided by Charlotte, the narrator in Victoria Glendinning's novel. Charlotte seems to be suggesting that all use of metaphor is conscious. But, as Nietzsche argues, truth and 'common sense' are what remain when self-consciousness about metaphor has disappeared. I then turn to economic theory to see both how metaphors can shape our view of the world and what happens when conscious reflection about one's choice of metaphor is no longer present. The discussion of economic theory has another purpose, related to the market metaphors found in many discussions of the Internet. The final section examines metaphors about the Internet, drawing primarily on the fifth anniversary edition of *Wired* and on policy discussions from the US. I shall do this in order to explore what they suggest about the normative order of the future and how we might arrive there, focusing particularly on their universalist claims.

What is a Metaphor?

In the extract from *Electricity* which introduces this paper, Charlotte describes a metaphor as the description of one thing in terms of another, to create a dramatic effect or because one cannot find any other way. Four explicit metaphors can be found in that extract: electricity as fluids, the universe as power station, supernatural omniscient being as human emotions and animals as other supernatural beings (the latter two draw on and reinforce the dualism of human/animal and good/evil). She also introduces the future metaphorical role of electricity as sexual attraction, crucial for the unfolding of her narrative.[1] Metaphors are rhetorical figures of speech in two senses, both originating with Aristotle. The first includes all figures of speech which achieve their effect through association, comparison and resemblance. Antithesis, hyperbole, metonymy and simile are all types of metaphor. This is not the dominant usage at present, although this meaning underlies the way we contrast 'metaphorical' with 'literal'. The more common way in which we now use metaphor is, as Charlotte explains, as a figure of speech which compares two things by saying that one is the other.[2] A theory compared to a building becomes a building, and as such acquires the following entailments: theories are constructed (developed) using building blocks (concepts and assumptions) to lay the foundations (evidence), and they can be undermined (subject to critique).

Aristotle's view of metaphor remains, but his additional suggestion that metaphor be used as a source of insight has not been developed within modern philosophy. Metaphor was relegated to the realm of emotion, imagination and subjectivity and was thus antithetical to reason, rationality and objectivity. Yet intellectual thought is nearly always guided by abstraction, in which 'reality' is expressed in terms of entities and their relationships to one another (Whitehead, 1926). Metaphor is an example of such an intellectual process, and as such it pervades our language and thinking. Lakoff and Johnson propose a way of moving beyond this particular dichotomy and argue that metaphor can unite imagination and reason.

> Reason ... involves categorization, entailment and inference. Imagination ... involves seeing one kind of thing in terms of another kind of thing... Metaphor is thus *imaginative rationality*. Since the categories of our everyday thought are largely metaphorical and our everyday reasoning involves metaphoric entailment and inferences, ordinary rationality is therefore imaginative by its very nature (Lakoff and Johnson, 1980 p.193, emphasis in original).

Lakoff and Johnson (1980) begin with Charlotte's definition of a metaphor, but go on to distinguish between three main types: structural, spatial or orientational, and ontological. Structural metaphors are those in which one concept or object is described in terms of another, such as electricity is a fluid. Mark Stefik (1996) presents a number of structural metaphors of the Internet: library, post-office, marketplace and 'other' worlds. Spatial metaphors are amongst the most pervasive in our language as they draw upon fundamental physical experiences such as up-down, in-out, front-back, near-far. Up and down, for example, are associated with emotional, physical, cognitive and moral states: 'she's on a high', 'wake up', 'we're engaged in high level intellectual debate', 'she has high standards.'

Our experience of ourselves and other entities is a rich source of the third type of metaphor identified by Lakoff and Johnson - ontological metaphors. This is reflected in our habit of attributing human qualities, especially agency, to non-humans, both animate and inanimate; and in our tendency to think of ourselves in terms of other entities. The Industrial Revolution spawned numerous machine metaphors: 'I'm a bit rusty', 'to let off steam.' Bernard Joerges suggests that the life-death metaphor organises much social science thinking about industrial technologies. For example, Karl Marx characterises human activity as 'living labour' and machine activity as 'dead labour'; Max Weber distinguishes the 'lifeless machine' of the factory from the 'living machine' of bureaucratic organisation; and Jürgen Habermas contrasts the 'system' with the 'life world' (1989 pp.31-2).

The development of computers has provided a rich new source of ontological metaphor, reflecting our attempts to understand these powerful machines and our own role in relation to them. The mind/body dualism is sometimes recast as one of software/hardware. Sherry Turkle observed that computers are the modern inkblot, projective devices for thinking about humans and social organisation. She suggests that children are increasingly defining themselves, 'not with respect to their differences from animals, but by how they differ from computers' (1984 p.313). Computers have memories and are susceptible to viruses and bugs. We now crash, suffer overload and run out of memory. These are all variations of the basic 'mind is machine' and 'computer is organic' metaphors. Jason Lanier, one of the contributors to *Wired*, writes, '[c]omputation has become the universal metaphor. The brain, the economy, evolution, and politics all feel like computer programs to an awful lot of people, even on the street' (1998 p.60).[3]

Metaphors in Economics

Before discussing emerging and competing metaphors of the Internet, I shall illustrate the cognitive and normative implications of using metaphors through an examination of economic thought. This is important not only for economic theory and policy but also for the next section because of the prevalence of market metaphors in discussions of the Internet.

Within classical economics, the dominant metaphors derive from the mechanical world view of Newtonian physics. Adam Smith's 'invisible hand' (1974 [1776]) and David Ricardo's image of the economic order as a gravitation process are examples of remote forces operating at a distance to maintain a system (1973 [1871]). Marx used biological metaphors in his discussions of socio-economic transitions in general and technological change in particular. On the whole, however, he rejected Darwinian theories of evolution because of their gradualism and emphasis on struggles for existence. Darwinism was not consistent with his vision of the class struggle, characterised by rupture and dialectical change. Joseph Schumpeter also deployed biological metaphors; his use of mutation as a descriptor of change, for example. He too rejected the Darwinian 'postulate that a nation, a civilisation, or even the whole of mankind (*sic*) must show some kind of uniform, unilinear development' (1934: p.57, cited in Clark and Juma, 1988: p.212).[4]

Alfred Marshall, one of the first neo-classical economists, also adopted some evolutionary metaphors for understanding the selection mechanisms at play in the growth and survival of firms (1997). His views about equilibrium, however, owe more to the laws of thermodynamics than to either Newtonian physics or Darwinian biology. Mainstream economic theory remains committed to the neo-classical model which emphasises short-term, static, equilibrium states. In this model, economic systems are understood as units of production (firms) and units of consumption (households) which exchange goods and services (including labour) in markets at prices which reflect the forces of supply and demand. Because of competition amongst both buyers and sellers, the price mechanism ensures that markets tend to equilibrium. This model of perfect competition requires that all economic actors have full information and respond rationally to changes in the cost of inputs. Neoclassical economics draws upon two metaphors central to capitalism: 'time is money' and 'labour is a resource'. Both of these reinforce the importance of time- and labour-saving technological change and contribute to peoples' alienation from their own labour. Time and labour can be made to fit equilibrium models of supply and demand which are regulated by price changes. Without wishing to deny the very real political and economic aspects of imperialism and globalisation, our understanding of these processes could be enriched by consideration of the imposition of metaphors developed in industrialised, capitalist societies on other parts of the world.[5]

Why does neo-classical economics, dominant within the economics profession, continue to adhere to models of equilibrium and stasis, especially when physics itself has largely abandoned them? The first possible answer is ideological, or normative. Individual greed, sanctioned by Smith's 'invisible hand' serves the *status quo* very well (1974 [1776]). The second reason is cognitive. Newtonian physics and thermodynamics validate a view of nature (and by metaphorical extension of the economy and society) in which discrete entities are linked together by different forces which are capable of self-regulation. This view works well for describing the behaviour of large, inert systems; it does not, however, work very well for explaining living systems of any size or complexity (Clark and Juma, 1988 p.214). There is a third reason which links these two. Perfect competition is the idealised system against which economic systems are judged. Even though economists know that 'reality' is characterised by numerous market imperfections (such as monopoly, uncertainty and imperfect information), they continue to promote policies that might move reality closer to the normative standard of perfect competition. Equilibrium models continue to set the metaphorical pace for economics, in orientational and ontological terms, to the detriment of both economic theory and policy and the lives of millions.

This very brief account of economic theory illustrates four important features of the use of metaphor. First, metaphors can assist us to think about new phenomena and new problems.[6] Second, metaphors can become solidified, and thus inhibit thought about new phenomena and new problems. Third, a good metaphor can alter our understanding of the world. Finally, metaphors are contestable and there are real political and cognitive issues at stake, as continuing debates about economic and environmental theories and policies demonstrate.[7]

Metaphors of the Internet and Cyberspace

In this section, I explore metaphors found in *Wired*, the monthly journal for evangelical Internet enthusiasts and in policy discussions in the US. A central tenet of this chapter is that our choice of metaphor reveals both what we think about today's reality and what we expect of tomorrow's; but I do not agree with Roger Burrows' (1997) endorsement of Mike Davis's (1992) suggestion to read Gibson as 'prefigurative social theory'. Burrows argues that it is difficult to disentangle the recursive relationship between cyberpunk novels and urban theory. Describing literature as social theory can be seen as an attempt by social science to colonise literature. The latter may well provide more profound and more elegantly expressed insights into the human condition than does social theory; nonetheless, literature has its own norms, standards and objectives, not all of which are shared with social science. Nonetheless, I do agree with Burrows' conclusion that we should (re)read cyberpunk novels, not as social and political theory as he suggests, but as sources of metaphor upon which social actors can and do draw.

Gibson is usually credited with introducing the term 'cyberspace' in *Neuromancer*, in which the following description of a children's television programme appears.

> 'The matrix has its roots in primitive arcade games,' said the voice-over, 'in early graphics programs and military experimentation with cranial jacks.' On the Sony, a two-dimensional space war faded behind a forest

of mathematically generated ferns, demonstrating the spatial possibilities of logarithmic spirals; cold blue military footage burned through, lab animals wired into test systems, helmets feeding into fire control circuits of tanks and war planes. 'Cyberspace. A consensual hallucination experienced daily by billions of legitimate operators, in every nation, by children being taught mathematical concepts... A graphic representation of data abstracted from the banks of every computer in the human system. Unthinkable complexity. Lines of light ranged in the nonspace of the mind, clusters and constellations of data. Like city lights, receding...' (Gibson, 1993 p.67).

Fifteen years after its initial publication, this remains one of the best definitions of cyberspace: a consensual hallucination where we keep our money, talk on the telephone and manipulate digital symbols for a variety of purposes. It is a description which reminds us of the military origins and the popular application of the techniques to games, contributing to the emergence of the 'military-entertainment complex'. Although Gibson wrote this before what we now call the Internet had spread much beyond the military, the academy and very big business, it is a description which still resonates. Gibsonian cyberspace simultaneously deploys the orderly metaphor of a matrix with the chaotic image of the city.

Metaphors found in *Wired* do not always allow for the same ambiguity that Gibson suggests is characteristic of cyberspace. In the lead editorial, Louis Rossetto, the editor, reflects on what motivated the launch of *Wired* in 1993.

> What we were dreaming about was profound global transformation. We wanted to tell the story of the companies, the ideas, and especially the people making the Digital Revolution... .
>
> After a century of war, oppression, and ecological degradation, we've entered a period of peace, increasing prosperity, an improving environment, and greater freedom for a growing proportion of the planet (Rossetto, 1998 p.20).

Stephen Graham and Alessandro Aurigi suggest that: '[M]uch of the current hype and hyperbole surrounding the Internet and "Information Superhighway" rests on the utopian assertion that such networks will inevitably emerge to be equitable, democratic and dominated by a culture of public space, enrolling multiple identities into new types of collective, interactive discourse and "electronic democracy".' (1997 p.20) Within the burgeoning literature about the Internet and cyberspace, two alternative visions can be found. The first vision focuses on the emancipatory potential of the Internet, a technology which allows individuals to transcend the limitations of space, time and biology in order to forge new identities and communities with like-minded people across the globe. The second vision is the dystopian antithesis of the first. Instead of liberating individuals, the Internet becomes the focus of alienation - of people from their families and friends in their local environments, of information workers from their own labour and that of their colleagues - and it is the source of concerns about the proliferation of pornography and racism. It is important to move beyond this dualistic thinking and to recognise that the Internet may be associated with both positive and negative outcomes (see for example, Jordan, 1999a). In this chapter, I focus on the utopian vision, actively asserted by the contributors to *Wired,* and contrast it with the engineering metaphor of the 'superhighway'.

Contributors to *Wired* recognise the importance of metaphors. Virginia Postrel attacks the engineering metaphors of highways and bridges used by politicians, suggesting they carry with them the entailments of government funding, teams of experts and large bureaucracy.

> Like an earlier Clinton/Gore plan to overlay the Net with a centrally planned and federally funded information superhighway, their bridge to the future isn't as neutral as it appears. It carries important ideas: The future must be brought under control, managed, and planned - preferably by 'experts'. It cannot simply evolve. The future must be predictable and uniform: We will go from point A to point B with no deviations. A bridge to the future is not an empty cliché. It represents technocracy, the rule of experts (Postrel, 1998 p.52).

Gore's metaphor of the superhighway guided the development and implementation of a range of policies around the 'national information infrastructure' during the first Clinton administration (1992-96). The metaphor has been significantly more successful than the policies, reflecting the promises of freedom and mobility previously delivered by the car earlier in the twentieth century. Gore elaborated the metaphor in a press briefing in December 1993.

> One helpful way is to think of the National Information Infrastructure as a network of highways much like the Instates begun in the 1950s. These are highways carrying information rather than people or goods.... . I mean a collection of Interstates and feeder roads made up of different materials in the same ways that roads can be concrete or macadam - or gravel. Some highways will be made up of fibre optics. Others will be built of coaxial or wireless. But - a key point - they must be and will be two way roads (Gore, 1993, cited in Kubicek and Dutton, 1997 p.12).[8]

The engineering metaphor refers back to the metaphors of computing as utility, akin to electricity and transport. Such utility metaphors were more common in the 1970s (Abbate, 1994) and were used to help construct fast and reliable networks as well as to promote models of control and regulation common in 'natural monopolies' at that time. Postrel argues that, '[d]ynamists [contributors to *Wired*, for example] typically are drawn toward organic metaphors, symbols of unpredictable growth and change' (1998 p.54). She later suggests that, '[t]hey [dynamists] see markets not as conspiracies, but as discovery processes, coordinating dispersed knowledge' (1998 p.56). Postrel is unaware of the contradiction inherent in holding both organic and market metaphors simultaneously. As we saw earlier, metaphors drawn from neoclassical economics carry with them the stasis of eighteenth and nineteenth century physics.

Between the covers of *Wired*,[9] six overlapping metaphorical themes can be found: revolution, evolution, salvation, progress, universalism and the 'American dream'. Revolutionary fervour is sometimes mixed with religious imagery. George Gilder reminds us of the book of Genesis: 'In the beginning was the word - the code - and it is not reducible to anything else' (1998 p.42). Randall Rothenberg mixes religious imagery with highways in order to discuss markets: 'The Net ... is the highway leading marketers to their Holy Grail: single-sourcing technology that can definitively tie the information consumers perceive to the purchases they make' (1998 p.76). This marketing Holy Grail can only be reached because of the omniscient facilities of surveillance technologies. In his analysis of *Time* magazine, Stahl (1995) identifies the prevalence of magical metaphors in its coverage of computer technology during the 1980s. He suggests that although the predominance of such metaphors declined throughout that period, they nonetheless served to bolster the power and authority of computer entrepreneurs. Explicitly magical language has declined, although religious metaphors continue to appear.

Metaphors of revolution also appear frequently. Po Bronson describes what is happening in Silicon Valley.

> I explained how there used to be this ethos through Silicon Valley that everyone was on a mission to transform our society, not just with personal computers - the ultimate populist tool - but by creating decentralised models for the workplace and new religions based on self-enlightenment rather than church scriptures. We wanted to shake up the world. Ten, 15 years ago - people felt this call to arms. I told him about the skull-and-crossbones flag flown over Apple during the development of the Macintosh (Bronson, 1998 p.112).

One of the people Bronson interviews designs telephony software. He exhibits some weariness with the constant change of the digital revolution: 'I've got a friend who's 24, and he's at his fourth start-up. How many revolutions can you join? It's like Monty Python's *Life of Brian*: you can't keep straight the People's Front of Judea from the Judean People's Front ' (1998 p.110). Such weariness is rarely found between the covers of *Wired*.

Evolutionary metaphors are the most common, and often also carry images of progress and salvation. John Perry Barlow, co-founder of the Electronic Frontier Foundation, strongly asserts the evolutionary nature of the Internet in his keynote address to the 1994 USENIX conference.

> [I]t's far more useful to look at the development of the Internet in biological rather than structural terms. The Internet seems to me very much like a life form. It has all those characteristics. It is self-organizing. It adapts itself readily into the possibilities faced that it finds (Barlow, 1994 p.2).

Four short examples of evolutionary metaphors from *Wired* are given below:

> There is no global village... A village is stable; everyone knows his or her role. What's happened instead is that everything has become more fluid... Corporations are transnational, merging and splitting like slime moulds (Lanier, 1998 p.62).

> [T]he concept of evolution argues [sic] that - in the absence of an unimaginably huge alteration in the physical world, such as climate change or planet collision - humanity will continue to go forward... We ride the greatest trend of all (Simon, 1998 p.68).

> Like some kind of technological Godzilla, IP [internet protocol] has gobbled up WANs [wide area networks] and LANs [local area networks], leaving behind a trail of dying equipment vendors. ... And - whomp! - the IP snowball rolls on (Steinberg, 1998 p.80).

> Eat or be eaten. Even the little guys, the very little guys who are doing something very cool and important - the four-guys-in-a-garage start-ups are playing the acquisition game (Bronson, 1998 p.108).

Leaving aside the mixed metaphor of Godzilla and the snowball and the attribution of voice to the concept of evolution, the repeated invocations of evolutionary change, progress and salvation require more careful scrutiny. Recall the caution with which Marx and Schumpeter treated evolutionary theory. They rejected what they perceived as its inevitability and universalism. The contributors to *Wired* deploy evolutionary metaphors while at the same time they invoke images of revolution, of massive social change towards a society characterised by greater freedom and progress. What are the agents of this revolution? It seems to be a mixture of the market and the technology. Yet, as we have seen, the market within capitalism is meant to operate in accordance with models of static equilibrium. The technology, as Gibson recognised, is not neutral. It is largely the product of military research applied to the lucrative markets of games and entertainment. The reasons for optimism about a dynamic and egalitarian future would seem to be misplaced. Presenting technology as the asocial mechanism for emancipation removes people from the historical process of change, which might occur in different ways in different places.

Langdon Winner raised similar concerns in 'Mythinformation', published originally in 1984, during what was then more commonly called the 'computer' or 'microelectronics revolution'.

> [T]he same society now said to be undergoing a computer revolution has long since gotten used to 'revolutions' in laundry detergents, underarm deodorants, floor waxes, and other consumer products. ... Those who employ [revolution] to talk about computers and society, however, appear to be making much more serious claims. They offer a powerful metaphor, one that invites us to compare the kind of disruptions seen in political revolutions to the changes we see happening around computer information systems (Winner, 1986 p.99).

Winner invites the reader to consider the goals of the putative computer revolution and how they might contribute to greater social justice. One of the traditional claims of political revolutions concerns universal rights: to land, education, the democratic process, the means of production, for example. I shall now examine the claims to universalism implicit in the metaphors of the 'information revolution'.

Internet enthusiasts often claim that connection is a global process, albeit an uneven one. This is not unique to the Internet. Similar claims can be found in much literature and policy about industrialisation and modernisation more generally. Individuals, regions, nations will 'catch up'; those who are not connected now, will or should be soon. This is the real annihilation of space by time: the assumption that the entire globe shares a single time line of 'development', in which some groups are further ahead than others along this shared path.

John Perry Barlow is committed to the emancipatory potential of the Internet (Jordan, 1999b). He reports on his visit to Africa, where he went to test his optimism about its potential to, 'proceed directly from the agricultural epoch into an information economy' (1998 p.143). He took with him fifteen pounds of solar panels, two 3400 Apple PowerBooks, a Newton 2000 MessagePad, a Jaz drive, five incompatible transformer bricks and a large bag of power and telecom adapters. He remains optimistic, not least because of what he perceives to be the, 'overlap between the ability to make music - one of Africa's prowesses - and the ability to make code' (1998 p.158). '[A]ll this suddenly melds into a vision of a prosperous Africa of small towns and rural communities, networked to the global grid through a web of wires and hearts opened wider with oestrogen' (1998 p.156). Women are central to his vision of the future, arising from what Barlow perceives to be women's greater capacities for work and lateral thinking. He observed that women effectively ran both the agricultural and information economies in the African countries he visited. His optimism is thus partly based on an essentialist view of the talents and capacities of black people and women. Barlow downplays the transient inconveniences some people will experience. 'Will there be data sweatshops? Probably. But, just as the sweatshops of New York were a way station for families whose progeny are now on Long Island, so, too, will these pass' (1998 p.158).

Conclusion

Metaphors about the Internet abound. Much is at stake: namely, the design, use and control of a global communication infrastructure which has the capacity to transmit data, speech, sound and images in a variety of configurations for many different purposes. This chapter has only scratched the surface, but it has illustrated how metaphors can influence public debate, policy and theory. It has not explored the role of metaphor in guiding the design process (see, for example, Mambrey and Tepper, 1998), nor has it explored the fascinating metaphorical associations of 'home pages' and 'frontiers' (see Neice, forthcoming and Turner, 1999). What this chapter has focused on is threefold. First, it has demonstrated the recursive relationship between science, technology and policy on the one hand and language and metaphor on the other. Second, through an examination of metaphors in economics, it has illustrated the contradictions inherent in the biological and market metaphors deployed by Internet enthusiasts such as contributors to *Wired* magazine. Third, it has challenged the universalist claims made by those same enthusiasts for the inclusive potential of the Internet, and invited the reader to consider both whether the policy discourses surrounding the development of the technology really promote greater social inclusion and whether it is desirable to promote a single, globalising technological development.

Metaphors can be used by everyone to help us think about the future, but they are also one of the resources deployed by a variety of actors in order to shape the future. Thus, it is important to continue to monitor the metaphors at work in order to understand exactly what work they are doing to assist different social actors to build the future.

Acknowledgements

The work on which this chapter is based is supported by the *Virtual Society?* Programme of the Economic and Social Research Council, grant no. L132251050. Versions of this paper have been presented at two workshops: 'Urban Futures/ Technological Futures', Durham, England, April 1998 and 'Politics of Technology', Maastricht, The Netherlands, May 1998. I am grateful to the organisers and participants of both for the opportunity to present these ideas and for the comments and feedback. I am also grateful to Tim Jordan, David Neice, Hans Radder, Jon Turney and the editors of this volume for comments on an earlier written draft. Mistakes and omissions remain my own.

Notes

1. Similarly, chemistry entered the metaphorical lexicon for describing human relationships in the eighteenth century.
2. In abstract terms, A is to B as X is to Y is used to make A is to Y as X is to B. For example, life is to old age as day is to evening is used to talk about the evening of life. Richards (1936) attaches the labels, tenor, vehicle and ground to the different parts of a metaphor, thus providing us with a good example of a mixed metaphor. In the example above, life is the tenor, day is the vehicle and time is the ground. In our understanding of electricity, it is the tenor, fluid is the vehicle and flow or current is the ground. By the end of the extract, sexual attraction becomes the tenor and electricity the vehicle. Sometimes a metaphor is defined as the vehicle alone; sometimes the combination of tenor and vehicle and sometimes all three.
3. There are, of course, alternative metaphors of the mind: a sponge which soaks up information, an empty vessel to be filled, a Swiss army knife. Each of these metaphors can be found in neo-Darwinian accounts of human evolution as well as in theories of pedagogy and human cognition and learning, with concomitant theoretical and policy implications. Whether one accepts or rejects these types of metaphor reflects one's position regarding the nature of human consciousness: is the human brain simply a powerful computational device where neurons are on or off, or not?
4. In their introduction to *Biology as Society, Society as Biology*, Maasen et al. (1995) suggest that concern about the use of biological metaphors arises from their use by latter day eugenicists or present day racists.
5. Contemporary evolutionary and institutional economists, such as Christopher Freeman, Carlota Perez, Luc Soete and others, have revived the more organic metaphors

occasionally to be found in Marx and Schumpeter in order to develop economic theories which they argue are better able to explain the dynamics of both technological and economic change. This approach to economics also has roots in the Cambridge school, especially the work of John Maynard Keynes (1973 [1936]) who focused on the problems of disequilibrium, in particular on the problems associated with the under-employment of resources, especially the under-employment of labour.

6.   See Miller (1996) for an interesting realist account of the role of metaphor in scientific creativity, especially in physics. He argues that metaphors are an essential part of scientific creativity because they assist scientists to move from descriptions of the unknown to literal descriptions - scientific theories - of the world around us.

7.   See Hajer (1995) for a discussion of the role of metaphors in environmental policy and discourse.

8.   Kubicek and Dutton (1997) provide the URL for the full text. However, in November 1999 it is no longer there. Instead, one finds extensive descriptions of IT^2 (Information Technology for the Twenty First Century), in which President Clinton plays an important role and Vice-President Gore has disappeared. Similarly, the highway metaphor has disappeared, to be replaced by more literal discussions of infrastructure. The popular discourse in the late 1990s is dominated more by discussions of applications (see http://www.hpcc.gov/).

9.   At a public lecture at the Royal Festival Hall, London on 21 February 1998, John Browning, European contributing editor of *Wired*, suggested that its designers wanted it to look, 'as if it had dropped from the future'. The fifth anniversary issue is not atypical: it is dayglo orange with the aphorism 'change is good' superimposed in another shade of dayglo orange. To me, it looks as if it has been unearthed from the 1960s. I am unsure as to whether this is more revealing of my age or of the age of the designers of *Wired*.

References

Abbate, J. (1994) *Analogy is Destiny: The Role of Metaphor in Defining a New Technology*, presented at MEPHISTOS conference, February.

Barlow, J.P. (1994) *Stopping the Information Railroad*. Available at http://www.eff.org/pub/Publications/John_Perry_Barlow/HTML/info_railroad_usenix.h tml (16 November 1999).

Barlow, J.P. (1998) Africa Rising, *Wired*, January, 142-58.

Breazeale, D. (ed. and trans) (1979) *Philosophy and Truth, Selections from Nietzsche's Notebooks of the Early 1870s*, Humanities Press International, New Jersey.

Bronson, P. (1998) Is the Revolution Over?, *Wired*, January, 98-112.

Burrows, R. (1997) Virtual culture, urban social polarisation and social science fiction, in B. Loader (ed.) *The Governance of Cyberspace*, Routledge, London 38-45.

Clark, N. and Juma, C. (1988) Evolutionary theories in economic thought, in G. Dosi et al (eds) *Technical Change and Economic Theory*, Pinter, London 197-218.

Davis, M. (1990) *City of Quartz*, Vintage, London.

Eagleton, T. (1997) Spaced Out, review of 'Justice, Nature and the Geography of Difference' by David Harvey, *London Review of Books*, 24 April, 22-3.

Gibson, W. (1984) *Neuromancer*, Victor Gollancz, London (page references in this chapter are to the 1993 Harper Collins paperback edition).

Gilder, G. (1998) Happy Birthday Wired, *Wired*, January, 40-2.

Glendinning, V. (1996) *Electricity*, Hutchison, London.

Graham, S. and Aurigi, A. (1997) Urbanising cyberspace? The nature and potential of the virtual cities movement, *City*, 7, May, 18-38.

Hajer, M.A. (1995) *The Politics of Environmental Discourse, Ecological Modernization and the Policy Process*, Oxford University Press, Oxford.

Joerges, B. (1989) Romancing the Machine - Reflections on the Social Scientific Construction of Computer Reality, *International Studies of Management and Organization*, 19, 4, 24-50.

Jordan, T. (1999a) *Cyberpower: The culture and politics of cyberspace and the Internet*, Routledge, London.

Jordan, T. (1999b) New Space, New Politics? Cyberpolitics and the Electronic Frontier Foundation, in T. Jordan and A. Lent (eds) *Storming the Millennium: The New Politics of Change*, Lawrence and Wishart, London.

Keynes, J.M. (1973) *The General Theory of Employment, Interest and Money*, Macmillan, London (first published in 1936).

Kubicek, H. and Dutton, W.H. (1997) The Social Shaping of Information Superhighways: An Introduction, in H. Kubicek, W.H. Dutton and R. Williams (eds), *The Social Shaping of Information Superhighways, European and American Roads to the Information Society*, Campus Verlag, Frankfurt/New York, 9-44.

Lakoff, G. and Johnson, M. (1980) *Metaphors We Live By*, University of Chicago Press, Chicago.

Lanier, J. (1998) Taking Stock, *Wired*, January, 60-2.

Maasen, S., Mendelsohn, E. and Weingart, P. (1995) Metaphors: Is there a bridge over troubled waters?, in S. Maasen, et al. (eds), *Biology as Society, Society as Biology*, Kluwer, Amsterdam, 1-8.

Mambrey, P. and Tepper, A. (1998) Technology Assessment as Metaphor Assessment – Visions guiding the development of information and communications technologies'. Presented at the European Association for the Study of Science and Technology conference, Lisbon, Oct.

Marshall, A. (1997) *The Principles of Economics*, Prometheus, New York (first published in 1890).

McCloskey, D. (1986) *The Rhetoric of Economics*, University of Wisconsin Press, Madison.

Miller, A.I. (1996) *Insights of Genius: Imagery and Creativity in Science and Art*, Springer, New York.

Neice, D.C. (forthcoming) Cyberspace and Social Distinctions: Two Metaphors and a Theory in R.E. Mansell (ed.) Inside the Communication Revolution – New Patterns of Social and Technical Intermediation, Oxford University Press, Oxford.

Postrel, V. (1998) Technocracy R.I.P., *Wired*, January, 52-6.

Ricardo, D. (1973) *The Principles of Political Economy and Taxation*, Everyman, London (first published in 1817).

Richards, I.A. (1936) *The Philosophy of Rhetoric*, Oxford University Press, Oxford.

Rossetto, L. (1998) Some things never change, *Wired*, January, 20.

Rothenberg, R. (1998) Bye-Bye, *Wired*, January, 72-6.

Simon, J. (1998) The Five Greatest Years for Humanity, *Wired*, January, 66-8.

Smith, A. (1974) *The Wealth of Nations*, Pelican, London (first published in 1776).

Stahl, W.A. (1995) Venerating the Black Box: Magic in Media Discourse on Technology, *Science, Technology and Human Values*, 20, 2, 234-58.

Stefik, M. (ed.), (1996) *Internet Dreams: Archetypes, Myths and Metaphors*, MIT Press, Cambridge, MA.

Steinberg, S.G. (1998) Schumpeter's Lesson, *Wired*, January, 80-4.

Thomas, G. and Wyatt, S. (1999) Shaping Cyberspace – interpreting and transforming the Internet, *Research Policy*, 28, 681-98.

Turkle, S. (1984) *The Second Self,* Simon and Schuster, New York.

Turner, F. (1999) *Cyberspace as the New Frontier? Mapping the shifting boundaries of the network society,* Red Rock Eater News Service. Online posting. Available e-mail: rre@lists.gseis.ucla.edu (6 June).

Whitehead, A.N. (1926) *Science and the Modern World,* Free Association Books, London.

Winner, L. (1986) *The Whale and the Reactor, A Search for Limits in an Age of High Technology,* University of Chicago Press, Chicago.

Part Three

Past Futures

7 Lessons from Failed Technology Futures: Potholes in the Road to the Future

FRANK W. GEELS AND WIM A. SMIT

Introduction

New technologies and their impact on society have always been the subject of speculation.[1] And technological developments are always accompanied by images of the future. This goes for the early stages of new technological developments, when people speculate about the eventual shape of the technology and its embedment in society. But it also goes for established technologies, when people expect the evolution in current technological trajectories to continue forever. Images of future development, together with parallel assessments of the current situation, abound in the implicit and explicit strategies that guide firms, government investment and user purchases.

Future images of the development and impact of technology can often be seen to have gone unrealised when judged retrospectively. Technological developments take different directions than predicted, their speed of development is slower or faster than imagined, or certain impacts on society are not foreseen at all. Apparently, there are potholes in the road to the future. This chapter asks the question why certain future images of technological development and impact do not materialise. In turn, when we understand the reasons why future images fail it may be possible to identify pitfalls and formulate lessons.

This question has relevance for technology policy as well, for example, in choosing which technologies to support, and in allocating resources to different technologies. Such choices are actually guided by expectations and images about future technological developments and how these will shape society. In better understanding the factors that contribute to the failure of future images, it may be possible to formulate a more reflexive technology policy.

Our approach here is to look mainly at the *content* of future images. More specifically, we look at the way technological developments and their impacts on society have been conceptualised. Upon close analysis, it transpires that many future images are based upon too simplistic conceptualisations of the 'impact' of technological developments on society. In particular, the neglect of the dynamic co-evolution of technology and society is an important cause for the failure of many future images.

We empirically investigate the past evolution of the predicted impact of information and communication technology (ICT) on traffic and transportation. Speculations on the implications of information and telecommunication technology (ICT) in this respect date back to the introduction of the telephone more than a century ago. Given this long history, adequate analysis of the images linking the future of ICTs to that of transportation necessitate a longitudinal. This is a particularly useful approach since many of the most pervasive images of the past in this domain have failed to materialise. Last but not least, the choice of topic has a more contemporary practical relevance since many Ministries of Transportation have high expectations of ICTs to alleviate, if not solve, transportation problems like congestion, air pollution and energy use.

The structure of the chapter is as follows. First, we discuss longitudinal shifts in the future expectations of the impact of ICTs on traffic and transportation (Geels and Smit, 1997a, b). Several descriptive methods are used, ranging from literal quotations to overview tables of quantitative estimations. We do not aim to present all past images in this chapter nor do we strive for a complete description of them but, instead, illustratively draw upon such sources to sufficiently support the general points of the argument. This analysis takes place in the subsequent section, where we seek to arrive at an answer to our central question: *why so many past images of the future fail?*

Using insights from technology studies, we distil a number of recurring key features from the empirical material, each of which imply conceptualisations or ways of thinking that may cause future images to fail. We then translate these key features into more general lessons and pitfalls for future speculations about the development and impact of technology. The lessons and pitfalls focus upon the content of speculations about the future, in particular upon the (implicit) conceptualisations of how technology develops and shapes society.

The factors that contribute to the failure of speculations may be interpreted as the result of ignorance or short-sightedness of the forecasters and futurists, lacking insights from technology studies. Although this interpretation often has validity, there is a complementary explanation for at least one of the key features. The penultimate section describes this complementary interpretation which is related to performative role of expectations in technological developments. In this interpretation, future speculations are identified as strategic resources in political and technological agenda-setting processes.

ICTs, Traffic and Transportation

Speculations on the impact of information and telecommunication technology (ICT) on traffic and transportation range back to the introduction of the telephone more than a century ago. Telephony's gradual introduction was accompanied by vehement discussions about its impact on social life. Although transportation itself was not so much an issue in these discussions, people expressed their fear that future generations might become isolated, maintaining necessary contacts only via the telephone. Fischer (1992, p.224) quotes the following future image from 1893, describing life in 1993:

> Families would live on scattered homesteads, neighboured only by people of like sentiment and quality, would conduct their work electronically, and would meet each other only at ceremonial occasions.

Thus, the current idea of an 'electronic cottage' is not as new as it may seem. Nor is the idea of the automated car, which was already part of popular thinking as early as the 1930s.

Table 7.1 gives some examples and preliminary indications about the time of their popularity. In order to present a more systematic description we make a distinction between two types of impact of ICT on transportation. *Direct* impacts refer to the implementation of information technologies in the transportation system itself. *Indirect* impacts refer to the use of information technology in business (e.g. working at distance from the office) or otherwise, that have a spin-off effect on transportation.

Table 7.1   Speculative impacts of ICT on traffic and transportation

Impact of ICT on traffic and transportation	Brief description
Direct impact	
Steering systems	Guidance of cars using computers; • 'feet off': computer controls gas pedal and velocity [popular in mid-1980s]. • 'hands off': computer and sensors steer the car in a limited fashion • 'brains off': computer steers the car to given destination [popular in late 1930s, mid-1960s and 1970s].
Informative systems	Board computer receives relevant information via traffic telematic systems, processes it and provides suggestions to driver. Driver uses information for decisions (e.g. about alternative routes or transport modes). [popular from mid-1980s].
Indirect impact	
Substitution of physical goods	Certain material flows (e.g. paper, news papers) are replaced by computers and telematics [popular in mid-1960-70s].
Teleconferencing	Communication via image-telephone or computer substitutes for business travel [popular in mid-1960s and 1970s].
Tele-working (-commuting)	People work on computer systems outside the office (e.g. at home), thus reducing commuter traffic [popular in 1970s and 1980s].

Direct Impacts

Let us consider these expectations in greater detail by distinguishing between three periods for past images of the *direct impact* of ICT on transportation we distinguish three periods. The first period dates back to the second half of 1930s and consists of sweeping speculations about automatic cars and road systems. General Motors in particular, most spectacularly articulated these future images at the World Fair of 1939 in New York. In its Futurama exhibition, GM guided visitors in motorised vehicles through scale models, demonstrations and films, showing a future transportation system where almost everything would operate automatically. After visitors had identified a destination and pushed the buttons on their vehicles, the automatic electronic system with its roads, invisible rails and radar systems would take them there.

The second period ranges from the mid 1960s and through the 1970s. The dominant characteristic of many expectations from that period is the role of the computer as a central planning and steering device. A number of nice illustrations of this way of thinking can be found in the book *The World in 1984: The Complete New Scientist Series*, edited by Nigel Calder (1964). Here, scientists and engineers were invited to express their ideas about the future. Bournonville, for instance, an engineer at the Compagnie des Machines in Paris, expected that 'twenty years from now, a machine may be used to direct traffic for a large city' (p.140). And Glanville, director of the Roads Research Department of Scientific and Industrial Research, wrote:

> By automatic control of large numbers of detectors and traffic signals, the computer will be continuously assessing the traffic position over an area and organising it to obtain the most efficient flow (p.189).

This would result in a 50% increase in road capacity, average speed and a decline in the number of road casualties. Illustrations can also be found in other countries, such as The Netherlands. In a publication by the Foundation of Future Images of Technology (STT), Professor Van der Burgt (1968 p.49) expressed his vision that a central computer could take care of the speed, direction and integration of cars in smooth traffic flows. The road capacity would thus be increased by 100%. And in *Transport in the Future*, a book edited by the Dutch transportation expert Hupkes (1970), a number of automated city traffic systems were described. One such system incorporated a complex system of rails with many stops on which small cabins react to travellers' push-button commands. The cabins' movements were based on electronic systems and a large central computer was expected to do the steering.

The third period starts in the mid-1980s, when the technological areas of telecommunication, mass-communication, data-communication and data-processing start to overlap, constituting the new field of telematics. From then on speculations about 'traffic telematics' emerge. Initially, the speculations often relate to partial applications, and gradually they focus more on an integrated system including: i) information collection (by sensors, satellites, cameras), ii) information processing (a central computer aggregating and integrating incoming partial messages) and iii) information transmission (via beacons next to the roads to board computers in cars).

Although the theme from the second period (automated vehicle guidance), remains alive, the emphasis in many future expectations has shifted to informative systems (see Table 7.1). The general claim in many of these expectations is that traffic telematics will substantially improve the efficiency of transportation. In addition to improved transportation comfort (e.g. convenient route planning that avoids traffic jams), reduced traffic congestion, better safety and emission standards are expected to follow.

On the basis of these optimistic expectations many private and governmental R&D projects on transport telematics have been initiated. For instance, in 1986, the multinational electronics company Philips started developing the CARIN-system (CAR Information and Navigation).[2] This system uses an on-board computer in the car, a positioning system and a special digital radio. The project leader, Thoone (1987 p.77) expressed the expectation that the wide application of traffic telematic systems might improve road efficiency by 20%. A similar estimation can be found in the Dutch Ministry of Transportation's first White Paper about *Telematics, Traffic and Transportation* (1990), stating that: '[i]t is thought that road capacity may be increased by 30% with the help of telematics' (p.19).

Despite all these optimistic promises, there are also critics who point to many uncertainties that lie ahead. For instance, the high costs of integrating electronic devices in the existing infrastructure may be a continuing problem. Also the efficiency gains in congestion and environmental emissions are still highly uncertain, especially when congestion occurs on roads for which there is no alternative (Elzen et al., 1996, p.54). After some years of working on the implementation of traffic telematics, the second and third White Papers by the Dutch Department of Transportation

(*Telematics, Traffic and Transportation* [1993] and *Progress Evaluation Telematics, Traffic and Transportation 1993-1995* [1996]) were much less optimistic about the quick and smooth introduction of traffic telematics. The second White Paper (1993) argues for a necessary shift from technology-push to a more demand oriented approach (p. 22), referring to the limited absorptive capacity of users, delay in the emergence of co-operation between suppliers, organisational bottlenecks in technology's introduction and unexpectedly high costs (p.24). It also mentioned that the initial expectations were too high (p.61). The third White Paper (1996) no longer mentions quantitative expectations or objectives but instead emphasises process goals, network-building, consultation, all of which would take considerably more time than once expected (p.6).

Comparing the second and third period, we stress three points. In contrast to the second period, the emphasis in the third period was on decentralisation, providing supportive information to independent car drivers who would have to make decisions themselves. Although central aggregation and processing of information does occur, the central role for ICTs was in decentralised onboard computers in individual cars. It was no longer thought feasible for traffic problems to be dealt with by centrally organised and (bureaucratically) controlled systems and large computers. Instead, citizens would be provided with sufficient information in order to enable them to make the right decisions. In a publication by the privately funded Foundation for Society and Enterprise (SMO, 1995) entitled *Traffic Chaos and Transportation Hunger: A Perspective on Mobility*, Professor Bovy describes the situation as follows:

> Decisions about the guidance of the optimal use of place and time are no matter for super traffic control centres. The intelligence is situated mainly low in the system, at the level of individual users. They themselves can best make decisions about optimal routes and times on the basis of up to date information about present and to be expected traffic circumstances [our translation] (pp.35-36).

Second, when governments and firms embarked on implementing telematics into the traffic system, they were confronted with practical difficulties that had not been mentioned in earlier more ambitious speculations. The societal embedding process of the technology went much slower than was expected. Third, compared to the second period, expectations for improvements of road efficiency have decreased in the third period. Table 7.2 summarises some findings.

Table 7.2 Estimations on the direct impact of ICT on road efficiency

Reference	Estimation of increased road efficiency
Glanville (1964)	50%
Van der Burgt (1968)	100%
Thoone (1987)	20%
1990 White Paper *Telematics, Traffic and Transportation*	30%
1990s literature	Increased attention to practical problems of societal embedding; decreased efficiency estimates

Indirect Impacts

Regarding the *indirect impact* of ICT we focus on three areas of impact: *tele-conferencing, substitution of goods,* and *tele-working.* Visions about widespread *tele-conferencing* became popular in the 1960s, when speculations about a technological fusion between telecommunication, data-communication and television emerged. Of course, the telephone has always been a technology to facilitate some sort of tele-conferencing between two persons, but now the issue was to have real-time business meetings in which persons scattered globally would participate. Again *The World in 1984* edited by Calder (1964), provides some illuminating examples. According to Barry, of Granada Television, the need for business travel will decline, because 'the simple device of telephone-plus-television will often make the fastest journey seem unnecessary' (p.159). And Pierce, a researcher at Bell Telephone Laboratories, states:

> I see by 1984 greatly extended data communication and improved telecommunication as a substitute for travel (p.153).

Clare, a researcher at Standard Telecommunication Laboratories, not only articulates a vision, but also prescribes the direction that technological development should take:

> In order to remove the need for the majority of the personal contacts at present necessary in most business operations and to provide the facilities that obviate gatherings at conferences, telecommunications must provide a form of high-definition colour television associated with high quality audio-channels: every blush and nuance needs to be accurately conveyed (p.155).

These examples show that tele-conferencing is often linked with the promise of a reduction in business travel. The idea that it actually *substitutes* for business travel is based on the assumption that business processes require a fixed and static number of social contacts: an increase in tele-conferencing, therefore, would automatically lead to a reduced need for travel. The possibility that tele-conferencing might *add* an *extra* mode of social engagement and thus increase the total number of contacts was not considered. Though the substitution potential of tele-conferencing figures in many images of the since the 1960s, there are only a few quantitative estimates of the effect (see also Table 7.3).

Table 7.3 Estimations of substitution potential of tele-conferencing

References	Estimation of substitution potential
German estimate from 1974/75 (Petersen, 1977)	66 % of business travel in Germany
Goddard and Morris (1976)	34 % of office meetings in UK and an additional 10 % using video-systems
Kraemer (1982)	20-30 % of business travel
European Conference of Ministers of Transport (ECMT, 1983)	15-25 % of business travel
Henckel et al. (1984)	26 % of business travel in Germany

A second area for which indirect effects of the emerging information society on transportation had been predicted, was in the *replacement of physical goods*. Especially paper products (e.g. newspapers, letters, books, library, mail) were expected to be substituted by electronic devices and computer systems. Although the so-called 'paperless society' has rarely explicitly been connected with solving transportation problems, we include this effect because it clearly illustrates the strength of the substitution idea. Since the image of a future information society emerged in the 1960s, visions of a paperless society have proliferated. Computers and terminals would penetrate both work and home practices, replacing paper products. Once again, Nigel Calder's *The World in 1984* (1964) illustrates this. Samuel, a researcher at the Watson Research Centre in New York, predicted that by 1984 almost everybody would have his/her own computer or terminal and be able to log in on central files where all information would be stored. As a result, 'Libraries for books will cease to exist in the more advanced countries' (p.145). This vision articulated with the more general idea of the time that all future information streams would be channelled through computers. Samuel further expected that 'all office paperwork will cease to exist in twenty years'. Pierce, a researcher at Bell Telephone Laboratories, foresaw a form of e-mail and thought it would strongly decrease regular mail:

> I expect that most of the sort of business letters which are now sent by airmail, and perhaps most business correspondence, will be sent electronically from machine-readable records (p.151).

According to Barry (Granada TV), newspapers and news distribution would change drastically:

> Web-offset and gravure methods of printing have set off changes in the newspaper technique that are about to affect radically the whole newspaper economy... In the area of mass-communication it is predictable that, in the long-term, the 'newspaper' of the future will be electronic, if indeed it can be called a newspaper in the presently accepted sense at all (pp.157-158).

Almost two decades later, the idea of newspapers being entirely substituted still reverberates in popular thought. *Megatrends*, by the futurologist J. Naisbitt, illustrates the substitution thesis (1982):

> At a certain point in time some American cities will contain so many computers that local newspapers will decide to stop publishing on the basis of increasingly expensive paper.

As these quotations indicate, many future expectations assume that a new technology will always *substitute* old technologies. With hindsight, we now know that newspapers, letters, and books exist *alongside* e-mail, data-communication, electronic journals and so on. What is more, taking a look at the printer in our department, one notices that the access to Internet results in the printer pouring out more paper than ever before. Rather than a paperless office, we have got a 'printing office'. Thus, old and new technologies can apparently co-exist.

The third area thought to have indirect impacts on traffic and transportation was and is *tele-working* (or tele-commuting as it is often called in American literature), implying that employees don't work in a central office, but at dispersed places (e.g. their homes). Still, the employees can work together, because they are connected via computers and telecommunication networks. One speculated advantage of tele-working is that it could reduce commuter traffic, even if employees work only part of the time at home.

In fact, some tele-working already existed *in practice* before it became part of wide ranging future expectations. In the 1960s, several companies practised forms of tele-working to reduce costs and increase flexibility. The software firm F-International (the F standing for 'freelance'), for instance, had created a pool of highly educated free-lance employees, to whom it forwarded assignments from customers. These employees were paid per task. Another early example is the Dutch National Giro Bank. When its two centres for encoding the hand written parts of cheques into a computer were faced with a shortage of labour (mainly young women) Giro Bank chose a decentralisation strategy. It created a number of geographically dispersed small centres, from where the encoded messages were fed into a large central computer via data communication.

In the 1970s the *idea* of tele-working and the changes with which it was associated became very popular with academics and futurists engaged in speculations on the 'post-industrial' and 'information society'. Rather than a solution to business problems, teleworking became linked to solving wider societal problems. After the oil crisis in 1973, for instance, tele-working was suggested as a means to save energy. A number of studies with some considerable influence were published including: *The*

Telecommunications-Transportation Trade-offs: Options for Tomorrow by Nilles et al. (1976) and *Technology Assessment of Telecommunications-Transportation Interactions*, from the Stanford Research Institute (1977). Of particular significance is the latter's exorbitantly high estimate that 50% all office jobs could be performed from home, and this has been consistently referred to in subsequent literatures. Although the 1970s showed very few empirical studies on teleworking *in practice*, many speculative publications flourished creating a kind of bandwagon effect.

In the 1980s, speculations about tele-working became even more widespread, in particular through popular works from authors like Toffler (1980) and Naisbitt (1982). Toffler coined the term 'electronic cottage' in his book *The Third Wave*. He estimated the potential of home tele-workers between 35% and 50% of the working population in 1990, and points not only to advantages of reduced mobility but also to greater societal stability, less stress, less temporary relationships and greater involvement in community life. In the 1980s more academic researchers felt a need to get a better grip on the phenomenon of tele-working, resulting in more quantitative estimations, more empirical research and more conceptual discussions on the concepts of 'information worker' and 'tele-working'. Still, many studies use different definitions, making it difficult to compare across them (see Table 7.4).

Since the mid-1980s, governments, stimulated by the promised societal advantages of tele-working, have encouraged experiments and co-operated with companies to this end. Though the number of tele-workers has increased, the growth is much slower than predicted. Becoming engaged in implementation experiments, rather than in wide speculations, governments and companies are confronted with practical and down-to-earth problems. For instance, employees that are willing to try tele-working at home find that their houses lack space to set up ergonomically acceptable workplaces. Employees discover that they miss the informal and social interactions with their colleagues. The fading distinction between work and private life results in psychological problems in the family. Employees feel that tele-working reduces their career opportunities, as they have less contact with their superiors. On the other hand, managers feel that they have less control over their physically absent employees.

Thus, it becomes clear that the societal embedding of tele-working does not proceed automatically and smoothly. Instead, all kinds of existing practices need to be adjusted and changed. Against this background, critical voices and articles emerge, making the point that earlier future speculations have paid too little attention to potential bottlenecks in the broader diffusion of tele-work (e.g. Forester, 1988; Fokkema, 1990; Forester, 1992).

Table 7.4 Some estimations of numbers or percentages of tele-workers

References	Estimations
Stanford Research Institute (1977)	50% of white collar jobs (in the high range scenario)
Toffler (1980)	35-50 % of working population in information societies
Magazine 'The Futurist' (June 1983)	10-15 million tele-workers in 1990 in the USA
Nilles (1984)	20 million tele-workers in 2000 in USA
Electronic Services Unlimited (1984)	20 % of working population in 1990 in USA
Dutch Organization for Applied Scientific Research (TNO) in 1984[3]	50 % of office functions in 2010 in The Netherlands, for part of the week
Nilles (1985)	10 million tele-workers in 1990 in USA
Miles et al. (1985)	10-15 % of British working population in 1990; 15-20 % in 2000
Dutch Telecom in 1986[4]	600,000-1,800,000 people tele-working part of the week in 1995 in The Netherlands (11 – 34% of working population)
Weijers and Weijers (1986)	40,000 – 80,000 tele-workers in The Netherlands in near future (0.8 – 1.6 % of working population)
Meijer et al. (1992)	A potential of 25-37 % of working population (1.3 – 1.92 million people) in The Netherlands for at least one day a week

Another critique of the tele-work promises outlined above comes from traffic experts (e.g. Salomon, 1985; Salomon, 1986; Nijkamp and Salomon, 1987). According to these experts the first order effect of substitution of commuter traffic may not be the only effect of tele-working. They point out that traffic should be seen as an integrated whole of several subsystems including recreation, social journeys, commuting and shopping. Accordingly, changes in one subsystem are likely to have implications for other subsystems. Thus, the reduction in commuting traffic may have second order effects such as an increase in recreational or social traffic. Tele-working, therefore, may well have traffic *generation* effects besides *substitution* effects.

Analysis: Key features in Images of the Future Role of New Technologies

In this section we distil a number of recurring key features from the historical future expectations described above.

A *first key feature* is that expectations may be biased by the broader cultural concerns of the time. Peoples' perception of the future is often coloured by their 'real time' cultural lens, their contemporary expectations, concerns and hopes. This, to some extent, may also explain why certain future images gain widespread popularity.

Thus, the popularity of speculations on automated cars and roads in the second half of the 1930s may be explained partly as a reaction to the first years of the 1930s, which were characterised by world wide economic depression and unemployment. The Futurama-exhibition of General Motors can be interpreted as the wider cultural promise of a better society through (electronic) technology, the promise of a smoothly functioning society, characterised by abundance and ease. To escape such biases one might design currently 'unlikely' scenarios next to those we think are most compelling or plausible today. Such practices have been a characteristic practice amongst large firms like Shell for instance (Wack, 1985a, b).

A *second key feature* is that sudden new trajectories in technological developments may trigger shifts in future images. The emergence of the PC and smaller workstations around 1980 opened up a new technological trajectory, wholly different from the traditional path towards ever-larger mainframe computers. It also opened up a new direction of expectations on the impact of computers on car traffic. Whereas in the mid 1960s and the

1970s future images referred to centralised computers that would steer and control traffic systems, such expectations gave way in the mid 1980s to decentralised onboard computers. Connected with the shift from central to decentralised control, came the shift from *steering* systems to *informative* systems. Whereas in the 'centralised steering system', the driver need only punch in one's destination, sitting back while the central computer does the work, in 'informative systems' the driver still has to make decisions on how to act, but now based on information that received from the system.

At the same time, this shift from centralised to decentralised control correlated with similar trends in *Zeitgeist,* away from the idea that governments could determine societal developments (without suggesting that this is either linear or culturally determined). This example, therefore, also illustrates the influence of broad cultural trends, i.e. the first key feature mentioned above.

A *third key feature* is that in speculations about the future role of a new technology is often phrased in terms of replacing or substituting the old technology, that is, in terms of winners and losers. In fact, it often happens that the old and new technologies will co-exist and service different markets and customer groups. For example, electronic information storage and transmission has not resulted in the expected 'paperless society'. By contrast, even more printed paper is being produced and instead of a 'paperless society' a 'printing office' has emerged. Hence, in addition to a (more modest) *substitution* effect, a *generation* effect is taking place.

One explanation of this thinking in terms of winners and losers, is that initially a new technology often is given meaning from existing *technological* contexts: the first car was called a horseless carriage (Flink, 1988); initially, the telephone was seen as a special kind of telegraph (Fischer, 1992), as was the first radio, sending messages between two points (Basalla, 1988). Therefore, it may easily happen that the new technology is expected to substitute for the old technology.

The reason why people initially interpret the emerging new technology in terms of the old technology is that all the potentials, properties and new meanings of the emerging technology are still to be found out. This takes place in, what has been called the 'co-evolution of technology and society' (Rip, 1995), or the process of a technology's 'societal embedding' (Green, 1992), when learning processes occur whereby the technology is further developed, more applications are 'discovered', and new meanings are attributed and articulated. So it takes time to struggle out of the grip of the old technology's meaning.

A *fourth key feature* is the neglect of the *generation* of new activities, by assuming that the pool of existing social practices and needs remains basically unchanged. From this assumption it may seem logical that the new technology will *substitute* certain social practices. For instance, many of the past expectations and quantitative estimates of the impact of teleconferencing on business travel implicitly assumed that business processes required a fixed number of social contacts. Consequently, it was expected that the increasing contacts via tele-conferencing would result in a decrease of contacts via business travel. However, we now know that the pool of social practices in business has both been extended and intensified.

The same may be true for the expectations that tele-working would reduce traffic because it would substitute for commuting. While there will be substitution effects, traffic experts have recently pointed out that tele-working might also intensify mobility in different parts of the traffic system, like social and recreational trips. It is possible that the teleworker will make more non-commuting trips in the extra 'free' time (e.g. see friends and shop, etc.). Furthermore, other members of the household may use the car when it is not used for commuting trips. Some evaluations of teleworking experiments indicate that these extra generated trips are approximately 80-90% of the substituted mobility (MuConsult, 1996). In a small-scale research project Van Reisen (1996) found that tele-working reduces 6% of commuting trips whilst generating a 4% increase in private trips. However, this generated mobility mostly falls outside rush hour, thus reducing congestion. Another traffic generating aspect often mentioned is that teleworking may contribute to the relocation of residences to more distant, 'green' locations ('urban sprawl'). The substitution of trips on tele-working days is thus partly offset by longer trips on commuting days. This effect may be related to the empirical finding that, over the last few decades, the average *time* that is spent on travelling per day remains constant, approximately one hour (Hupkes, 1977; Cerwenka, 1985). This means that savings in commuting time as a result of (technological) innovations are converted into longer commuting trips or journeys for other purposes. With regard to this discussion about the effects of teleworking, it should be noted, however, that large-scale and long-term empirical studies are still lacking.

Nevertheless, it is clear that there are both substitution and generation effects. Therefore, the fourth key feature has in that respect a similar implication as the third key feature. However, the underlying assumptions revealed by the third and fourth key factor that cause future images to go wrong are quite different.

A *fifth key feature* is what may be called *functional thinking*. The expectation is often voiced today that the phenomenon of 'tele-shopping' or 'virtual shopping' via Internet, would substitute for physical shopping, thus reducing the need for traffic. Such future images view the social activity of shopping as a purely functional activity, the purchase of required goods. Such purely functional thinking neglects other social and psychological aspects that are involved in the activity of shopping.

For instance, Weijers (1996 p.70) has investigated the social and cultural aspects that are seen to inhibit large-scale tele-shopping. He acknowledges that many functional purchases could be and are being done by tele-shopping, and that this, indeed, could reduce transportation for customers (though it would increase it for retailers that have to deliver the goods).[5] However, for 'social', non-functional shopping physical presence is and will remain a compelling part of the activity. The social component may include elderly people for whom consumer contexts are a primary source of social encounter or, more generally, where shopping is already a way of relaxing with friends and family. Gender differences still play a role with regard to tele-shopping. Many women seem to enjoy the social aspect of shopping, while many men seem to be involved in interactive technologies. Not surprisingly, it might be legitimately argued that many of those who have formulated high expectations about the future of tele-shopping were men.

The main point of this is that from a purely functional perspective the future of certain new technologies may look shining, because they promise to fulfil a certain function more effectively. The danger of such strong functional thinking is that social and psychological aspects are neglected. As a result, future images that are based on strong functional thinking may turn out to be too optimistic about the diffusion, use and expected effects of certain technologies.

Table 7.5 Pitfalls in future explorations of technological developments

Pitfalls	Neglected aspects	Examples
Expectations have cultural biases, reflecting current beliefs, hopes, fears	Cultural changes or alternative directions of technological development	Centralised steering computer traffic systems in 1960/70s. Decentralised informative telematic systems from mid-1980s
Focus on current technological trajectories	Alternative technological futures	Same as above
Potential of new technology is phrased in terms of substitution of old technology	Co-existence of old and new technologies Generation next to substitution	Replacement of paper products by computer systems (paperless society vs. printing office)
Pool of existing social practices is assumed to remain constant in spite of the introduction of a new technology	Co-evolution of technology and society Generation next to substitution	Tele-conferencing and business travels Tele-working and impact on both commuting and recreational traffic subsystems
Functional thinking	Other social needs	Tele-shopping and sociable shopping
Overestimation of speed of societal embedding of new technology	Practical problems and adjustment processes	Telematics in traffic and transport Tele-working
Initial promises in future expectations are too high and have to be scaled down in later periods	The role of promises in technological development	Transport telematics and improvements of road efficiency. Tele-conferencing and substitution of business travels Growth of tele-working

A Complementary Explanation: The Role of Expectations in Technological Development

The seven key features in explaining the failure of images of the future can be interpreted as resulting from ignorance and short-sightedness of forecasters or futurists, lacking insights from technology studies and using too simplistic assumptions about the impact of technology. Although this interpretation often has validity, it is, in our view, only part of an explanation for the seventh key feature (which relates to initial promises of a technology's impact that turn out to be too high and are scaled down in later periods). Below, we give a complementary interpretation of this seventh key feature, which focuses on the performative role of expectations and future images in technological developments. Future images are not neutral predictions at a distance from technological development. They *do* something that affects the direction and speed of developments. They may even play the role of an intervention in technological developments. Before we can make the reflexive point about the societal role of future images, we need a brief detour in technology studies and discuss the role of promises and expectations in technological developments.[6]

New technological options often emerge as 'hopeful monstrosities' (Mokyr, 1990), 'hopeful' because they have demonstrated that they can fulfil some societal function, but they are 'monstrous' because their performance characteristics are low. For example, when the telephone was invented in 1876, the caller would have to shout loudly with the recipient having to listen carefully. Cars with internal combustion engines in the 1890s were dangerous and dirty machines that frequently broke down, and performed not much better than alternatives like the electric and steam vehicles. The ENIAC computer, completed in 1946, weighed 30,000 kg and consisted of 48,000 thousand switches the frequently malfunctioned and had to be replaced (Nijholt and Van den Ende, 1994).

Because of their low performance characteristics, new technological options cannot immediately compete on the market. They first need to be nurtured and further developed. Innovators of new technologies try to create a 'protected space' in which they can improve their technologies, hopefully increasing the performance characteristics. In quasi-evolutionary theories of technological developments these protected spaces are called 'niches' (Van den Belt and Rip, 1987; Schot, 1991; Weber and Hoogma, 1998). A niche consists of a network of actors (e.g. funding organisations, technology developers) that share a belief in the future potential of a new technology and are willing to invest time and money in further development.

Developers of new technologies have to do a lot of work (e.g. lobbying, convincing people) in order to create a protected space. The crucial point in the context of this chapter is that promises or diffuse scenarios about the potential of future technologies are crucial resources in the creation of niches. Promises and diffuse scenarios are used to convince funding organisations to invest money and attract other engineers to join the development.

The role of promises, expectations and diffuse scenarios in technological development has been investigated theoretically and empirically by Van Lente (1993 p.187), resulting in the following model of technological development. When a new technological opportunity emerges, its protagonists formulate promises and diffuse scenarios about its future potential. When the spokespersons are successful, the promises and future images become accepted in relevant communities, e.g. amongst policy makers or engineers. This process is called political or technological agenda setting. Money and other resources become available for the development of the new technology, provided that the broad promises are translated into concrete requirements. Technical experts in technical communities and governmental departments translate the future scenarios into requirements for concrete work. Because these requirements provide search heuristics for the engineers, they also influence the direction of technological development directly. On the basis of these requirements, funding becomes available to establish protected spaces where concrete activities can take place for a certain period of time. Through concrete development activities people learn about the technological characteristics and potential applications. When the time schedule of the funding cycle runs out, the outcomes are assessed. Van Lente (1993) calls this entire process the 'promise-requirement cycle'.

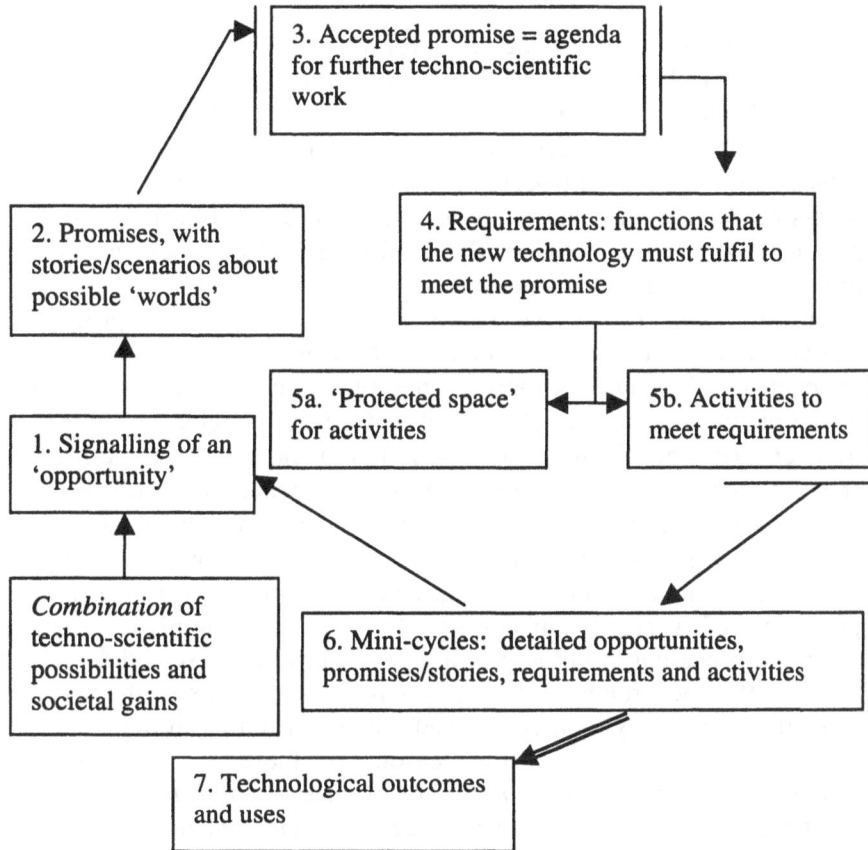

Fig. 7.1 The social function of promises and expectations in technological developments (Van Lente and Rip, 1998)

On the basis of the learning experiences, new and more specific opportunities will be identified again. Subsequently, a new cycle starts, and new promises are articulated that are more specific and detailed than the earlier ones. This may be repeated several times. From this perspective, technological development can be understood as consisting of a series of promise-requirement cycles. In Figure 7.1 we have schematically summarised the above discussion about the social role of promises and expectations.

The result of subsequent promise-requirement cycles is that technological characteristics, applications and user groups become more specifically defined and aligned. This is also true for the future expectations and promises that accompany technological development. During subsequent cycles a shift occurs from broad, diffuse expectations to future assessments that pay attention to the concrete applications and social practices involved. Put differently, a shift occurs from wide speculations to more practically oriented assessments.

In line with the above perspective on the role of promises and characteristics, we can now interpret the seventh key feature as a basic element in all kinds of new technological developments. During the development process, initially broad and diffuse expectations are replaced by ever more specific promises and requirements when people learn about specific technological characteristics and potential applications. For instance, while in the 1970s tele-working was based on an abstract *idea* and connected to socio-cultural issues such as the 'information society', most assessments in the 1990s take into account the down-to-earth practical problems of societal embedding. Indeed, the optimistic quantitative estimations present in initial promises often have to be scaled down in later periods. But their function was to convince other actors of the fact that the 'hopeful monstrosity' is indeed hopeful and worth investing in.

From this perspective, the reason for these too optimistic initial promises and expectations is *not* that forecasters or futurists are ignorant or short-sighted. Instead, promises are strategic resources in promise-requirement cycles. Initial promises are set high in order to attract attention from (financial) sponsors, to stimulate agenda-setting processes (both technical and political) and to build 'protected spaces'. Promises thus play a role in the social processes that are part of technological development. This performative dimension of future images provides a complementary interpretation of the failure of some future speculations. The seventh key feature (that is, initial promises are exaggerated and have to be scaled down in later periods) can be explained by the fact that promises and expectations are part of strategic games. Some future speculations do not strive for truth or accuracy, but are meant to influence specific social processes in technological developments.

Concluding Remarks

In this chapter we have sought to respond to the question why, in practice, certain future images of technological developments and their impact do not come true. Essentially, many failed future images are based upon too simplistic conceptualisations of the impact of technological developments on society. In particular, the neglect of the dynamic co-evolution of technology and society is an important cause for the failure of many future images. We have further specified this general cause into seven key features of the failure of future images, ranging from the neglect of wider cultural changes, to underestimation of practical adjustment processes in the adoption of technology. These key features are summarised in Table 7.5 in terms of pitfalls and neglected aspects. Our research also shows the relevance of technology studies for designing new future oriented methodologies, such as the development of socio-technical scenarios (Geels, 2000).

Insights from technology studies have also been useful in showing that there is a complementary explanation for at least one of the key features of failing technology futures. This explanation focuses on the performative role that promises and expectations play in the social process of technological development.

Does the phenomenon revealed by this complementary explanation about future expectations of new technologies constitute a serious problem? The answer is no and yes. No, because advocates of a new technology have to mobilise support and one resource for this are the promises that can be attached to a new technology. For this mobilising purpose, the advocates cannot do without some societal blinkers. Thus, it cannot be prevented that promises and expectations are part of strategic games. The answer is yes, when policy choices have to be made about which technologies should be supported and when decisions on investments have to be made. In that case policy makers should not go along too easily with promises of very high future impacts. Instead, they should have an eye for the pitfalls revealed by the seven key features (see Table 7.5). Accordingly, by contesting such futures in this way, it might become possible to engage in more reflexive decisions about the allocation of resources to new technological developments, taking into account a broader variety of possible societal impacts.

Notes

1. This chapter is based on research that has been funded by the Dutch Ministry of Transportation as part of the research programme 'The impact of the information society on traffic and transportation'. This programme was conducted in 1997, and consisted of five projects. The authors of this chapter were responsible for one project 'Future images in the past'. This has resulted in two reports (Geels and Smit, 1997a, b).
2. Other programmes are: The IntelligentVehicle Highway Systems (IVHS) in USA. Federal support amounted to $25 million in 1991/92 and $600 million for the five year period 1992-1996; Japan is leading in transport telematics. Since 1992 the ATIS-system (Automatic Transportation Information System) is for sale. It has become popular very fast and at the end of 1994 about 500.000 copies have been sold; and Europe stimulates co-operation between governments, car- and electronics industry with large scale R&D projects, e.g. PROMETHEUS, DRIVE I, DRIVE II.
3. Cited in report from Telematics Research Centre (1994).
4. Cited in Kok et al. (1993).
5. Logistic planning by the retailers may, however, still lead to a small net decrease in mobility.
6. The role of promises and expectations in technological developments fits into a sociological, quasi-evolutionary model of technological development, in which the generation of technological variety is interdependent with the (societal) process of selection of technologies (Van Lente, 1993).

References

Barry, G. (1964) Mass Communications in 1984, in N. Calder (ed.), *The World in 1984: The Complete New Scientist Series*, Penguin Books, Harmondsworth, 157-160.
Bassala, G. (1988) *The Evolution of Technology*, Cambridge University Press, Cambridge.
Bordewijk, J.L., Van den Berg, D. and Horn, W. (1973) *Communicatiestad 1985: Elektronische Communicatie met Huis En bedrijf*, Stichting Toekomstbeeld der Techniek, Koninklijk Instituut van Ingenieurs, The Hague.
Bournonville, L. de (1964) The Customers' Ideal Computer, in N. Calder (ed.), *The World in 1984: The Complete New Scientist Series*, Penguin Books, Harmondsworth, 139-142.
Cerwenka, P. (1985) Strukturwandel imMobilitätsbudget Durch Telecommunication, *Zeitschrift für Verkehrswissenschaft*, 55, 240-250.
Clare, J.D. (1964) Decentralization by Telecommunications, in N. Calder (ed.), *The World in 1984: The Complete New Scientist Series*, Penguin Books, Harmondsworth, 154-156.
Electronic Services Unlimited (1984) *Telecommuting Research Program*, ESU, New York.
Elzen, B., R. Hoogma and Schot J. (1996) *Mobiliteit met Toekomst: Naar een Vraaggericht Toekomstbeleid.*: Ministerie van Verkeer en Waterstaat, Directoraat-Generaal Rijkswaterstaat, Adviesdienst Verkeer en Vervoer, Rotterdam.
European Conference of Ministers of Transport (1983) *Transport and Telecommunications*, ECMT, Paris.
Fischer, C.S (1992) *America Calling: A Social History of the Telephone to 1940*, University of California Press, Berkeley.

Flink, J.J. (1988) *The Automobile Age*, MIT Press, Cambridge, Mass.

Fokkema, J. (1990) *Telewerken Dichterbij? Een Onderzoek Naar de Haalbaarheid van Werken met Telematica in de Woonomgeving*. Stichting Werkgroep 2 duizend, Amersfoort.

Forester, T (1988) The Myth of the Electronic Cottage, *Futures*, 6, 227-240.

Forester, T. (1992) Megatrends or Megamistakes? What ever Happened to the Information Society?, *The Information Society*, 8, 133-146.

Geels, F.W. and Smit, W.A. (1997a) *Toekomstbeelden in het Verleden: De Invloed van de Informatiemaatschappij op Verkeer en Vervoer*, Hoofdrapport, Rapport voor de Adviesdienst Verkeer en Vervoer, Rotterdam, Ministerie van Verkeer en Waterstaat, 54.

Geels, F.W., Smit, W.A. (1997b) *Toekomstbeelden in het Verleden: De Invloed van de Informatiemaatschappij op Verkeer en Vervoer*, Achtergrondrapport, Rapport voor de Adviesdienst Verkeer en Vervoer, Rotterdam, Ministerie van Verkeer en Waterstaat, 161.

Geels, F.W. (2000) Sociotechnical Scenarios as a Tool for Reflexive Technology Policies: Using Evolutionary Insights from Technology Studies, Sorensen, K. and Williams, R. (eds.), *Conceptual Advancements and Policy Implications of Recent Developments in Social Shaping Research*, Focused study on the Social Shaping of Technology, Report for the European Commission (forthcoming).

Glanville, W. (1964) Roads and Traffic in 1984, in N. Calder (ed.), *The World in 1984: The Complete New Scientist Series*, Penguin Books, Harmondsworth 187-189.

Goddard, J.B. and Morris D. (1976) The Communications Factor in Office Location, *Progress in Planning*, 6.

Green, K. (1992) Creating Demand for Biotechnology: Shaping Technologies and Markets, in R. Coombs, P. Saviotti and V. Walsh, *Technological Change and Company Strategies: Economic and Sociological Perspectives*, Academic Press Limited, San Diego, 164-184.

Henckel, D., Nopper, E. and Rauch, N. (1984) *Informationstechnologie und Stadtentwicklung*, Deutscher Gemeindeverlag, (Schriften des Deutschen Instituts fuer Urbanistik; Bd. 71), Stuttgart.

Hupkes, G. (1970) *Transport in de Toekomst: Koorddansen Tussen Wens en* werkelijkheid, Wetenschappelijke Uitgeverij N.V, Amsterdam.

Hupkes, G. (1977) *Gasgeven of Afremmen. Toekomstscenario's voor ons Vervoerssysteem*, Kluwer, Deventer.

Kok, A., Mante, E. and MeLieste, J. (1993) Telewerk, de weikvorm van de toekomst, *Studieblad PTT Telecom*, May, 268-293.

Kraemer, K.L. (1982) Telecommunications / transportation Substitution and Energy conservation, *Telecommunications Policy*, 7, 39-59.

Meijer, R.A.M., Wijers ,T.C.M., Spoelman, E.J. and Rip, A. (1992) *Telewerk Blijft Maatwerk: De Invoering van Telewerk op Grote Schaal*, TNO-Beleidsstudies, Studiecentrum voor Technologie en Beleid TNO, Apeldoorn.

Miles, I. (1985) *IT Futures Surveyed*, Innovation Research Group and Science Policy Research Unit, Brighton, Working Paper.

Ministerie van Verkeer en Waterstaat (1989-1990) *Telematica, Verkeer en Vervoer*, SDU-Uitgeverij, The Hague, Parliament Number, 21, 449.

Ministerie van Verkeer en Waterstaat (1993-1994) *Voortgangsnota Telematica, Verkeer en Vervoer*, SDU-Uitgeverij, Parliament number 22, 300, The Hague.

Ministerie van Verkeer en Waterstaat (1996), *Telematica Verkeer en Vervoer, Eindrapportage 1993-1995*. Coördinatiepunt Telematica, The Hague.

Mokyr, J. (1990) *The Lever of Riches*, Oxford University Press, New York.

MuConsult (1996) *De Thuisblijvende Auto: Analyses OVG 85-95.* Report for Ministry of Transportation, Rotterdam.

Naisbitt, J. (1982) *Megatrends. Ten New Directions Transforming our Lives*, Warner Books, New York.

Nijholt, A. and Van den Ende, J. (1994) *Geschiedenis van de Rekenkunst, van Kerfstok tot Computer*, Academic Service, Schoonhoven.

Nijkamp, P. and Salomon, I. (1987) Telecommunication and the Tyranny of Space, in I. Orishima, G.J.D. Hewings and P. Nijkamp (eds.), *Information Technology: Social and Spatial Perspectives*, Proceedings of an International Conference on Information Technology and its Impact on the Urban-environmental System, Held at Toyohashi University of Japan, November 1986, 91-104.

Nilles, J.M. (1984) *Managing Teleworking: A Project in the Information Technology Program*, Center for Futures Research, Los Angeles.

Nilles, J.M. (1985) Teleworking from home, in T. Forester (ed.), *The Information Technology Revolution*, Basil Blackwell Ltd, Oxford, 202-208.

Nilles, J.M., Carlson, T.R., Gray, P. and Hanneman, G.J. (1976) *The Telecommunications-Transportation Tradeoff: Options for Tomorrow*, John Wiley & Sons, New York.

Petersen, H. (1977) Telekommunikation und Verkehr, *Internationales Verkehrswesen*, 224-228.

Pierce, J.R. (1964) Private Television Instead of Travel, in N. Calder (ed.), *The World in 1984: The Complete New Scientist Series*, Penguin Books, Harmondsworth, 151-153.

Reisen, F. van (1996) Telewerk in Samenhang met Andere Strategieën ter Beheersing van Automobiliteit en Congestie, *Colloquium Vervoersplanologisch Speurwerk*, 2, 717-736.

Rip, A. (1995) Introduction of New Technology: Making Use of Recent Insights from Sociology and Economics of Technology, *Technology Analysis & Strategic Management*, 7, 417-431.

Salomon, I. (1985) Telecommunications and Travel: Substitution or Modified Mobility?, *Journal of Transport Economics and Policy*, 219-235.

Salomon, I. (1986) Telecommunications and Travel Relationships: A Review, *Transportation Research*, 20a, 3, 223-238.

Samuel, A.L. (1964) The Banishment of Paper-work, in N. Calder (ed.), *The World in 1984: The Complete New Scientist Series*, Penguin Books, Harmondsworth, 142-147.

Schot, J. (1991) *Maatschappelijke Sturing van Technische Ontwikkeling: Constructief Technology Assessment als Hedendaags Luddisme*, University of Twente Press, Enschede, PhD thesis.

Stanford Research Institute (1977) *Technology Assessment of Telecommunications-transportation Interactions.* Vol. I, II, III, Menlo Park, California.

Stichting Maatschappij en Onderneming (1995) *Verkeerschaos en Vervoershonger: Perspectief op Mobiliteit*, SMO, The Hague.

Telematics Research Centre (1994) *Telewerk, Werkbaar of Werkzaam? Twee Visies op de Toekomst*, Telematics Research Centre Publications, Enschede.

Thoone, M.G.L. (1987) Informatiesystemen voor Verkeer en Vervoer, *Informatie en Informatiebeleid*i, 5, 69-77.

Toffler, A. (1980) *The Third Wave*, Bantam Books, New York.

Van den Belt, H. and Rip, A. (1987) The Nelson-Winter-Dosi Model and the synthetic dye chemistry, in W.E. Bijker, T.P. Hughes and T. Pinch (eds.), *The Social Construction of Technological Systems: New Directions in the Sociology and History of Technology*, MIT Press, Cambridge, 159-90.

Van der Burgt, G.J. (1968) Toekomstige Ontwikkelingen in het Wegverkeer, in J.L.H. Cuperus (ed.), *Verkeersmiddelen*, Stichting Toekomstbeeld der Techniek, Koninklijk Instituut van Ingenieurs, The Hague, 41-54.

Van Lente, H. (1993) *Promising Technology: The Dynamics of Expectations in Technological Developments*. Eburon, Delft., PhD thesis University of Twente, Enschede.

Van Lente, H. and Rip, A. (1998) Expectations in Technological Developments: An Example of Prospective Structures to be Filled in with Agency, in C. Disco and B.J.R. Van der Meulen (eds.), *Getting new technologies together*, Walter de Gruyter, Berlin, 195-220.

Wack, P. (1985a) Scenarios: Uncharted waters ahead, *Harvard Business Review*, September-October, 73-89.

Wack P. (1985b) Scenarios: Shooting the rapids, *Harvard Business Review*, November-December, 139-150.

Weber, M. and Hoogma, R. (1998) Beyond National and Technological Styles of Innovation Diffusion: A Dynamic Perspective on Cases from Energy and Transportation Sectors, *Technology Analysis and Strategic Management*, 10, 545-566.

Weijers, T. and Weijers, S. (1986) *Telework: Een overzichtsstudie naar Recente Trends en Toekomstperspectieven*, Ministerie van Sociale Zaken en Werkgelegenheid, The Hague.

Weijers, S. (1996), *Mobiliteit en Telematica Voorzieningen, Meer Aanbod en Minder Vraag?*, Pb-IVVS, Den Haag.

8 Science Fiction's Memory of the Future

HILARY ROSE

As befits a literary genre invented not only by a woman but by a very young woman, for Mary Shelly's *Frankenstein* (1819) is unquestionably the foundational text of anglophone science fiction (SF), SF has never attained higher literary status. Yet paradoxically its very marginality has provided a cultural space for innovation relatively unpoliced by literary criticism. Thus when the distinguished novelist Doris Lessing (1979; 1980; 1981) made a foray into the genre, this part of her literary output was more or less ignored. However with the advent of cultural studies, with its distaste for dividing culture hierarchically, SF came under academic scrutiny with at least two journals devoted exclusively to the field. Thus from the *Times Literary Supplement* I learn that the reviewer considers the most recent issue of *Science Fiction Studies* as lucidly written and jargon free, further that it does not 'labour under the misapprehension that writing academic essays about science fiction is going to bring about a global revolution and inaugurate a new, non-hegemonic discourse' (Irwin, 2000 pp.50-56). Although it is difficult to entertain seriously the idea that anyone one actually thinks that an academic essay could be quite so miraculous in its effectiveness, I too want to stand up for SF (or rather some kinds of SF) as a laboratory for exploring alternative futures.

Literary critics interested in SF often suggest that the interwar period was a golden age for the genre. Those of us interested in social criticism of science, the radical science movement and indeed the founding years of what was to become science and technology studies, also see this period as a golden age. My argument is that this is not by chance, the discussion of science fed into a huge debate about not just economy and society but also culture, including science as culture. The sheer intensity and richness of the arguments and the writing indeed make this a golden age. But I want to argue that there was another golden age with the advent of second wave feminism and the creative explosion of feminist literary writing, especially

157

in the form of feminist SF (FSF). Again the critiques of science and technology being advanced by feminist science theorists, the profound political struggles to control technology (most of all those around reproductive freedom) and the new FSF constituted a heady conjunction.

My own interest in SF developed as part of my long-standing involvement both as an analyst of, and activist within, the radical science movement. While not unappreciative that even a weak social capitalism is a substantial improvement on a raw market model, to say nothing of anything more socially gruesome, I have never been able to sign up to the view that this is the best of all possible worlds. Thus while entirely without interest in the genre of SF for the same reasons argued below by Ursula Le Guin, I nonetheless read the futurist scientific writing of the twenties and thirties Marxist scientists simply because they had a progressive cultural and political agenda. Yet because reading them and becoming a feminist occurred almost at the same time in my personal biography, they became a fast entry point into understanding the limits of the Marxist agenda. Futurists' utopias have the advantage that they make wonderfully transparent the values that inform them. Culturally speaking they offer a quick litmus test, not least for issues of gender. The feminist laboratory of dreams and nightmares on offer in feminist science fiction gives energy and releases the imagination to that huge task of taking women out of nature and repositioning ourselves in culture. Arguably the extreme social constructions of some influential strands of feminism in those the early years generated other theoretical and practical problems but that is not my concern here. Here I want to insist on the importance of what Christa Wolf (1977) speaks of as 'the memory of the future', for that is the heart of a politically and culturally engaged SF.[1]

The Founding Teenage Mother of Science Fiction

Let us begin by returning to the founding teenage mother of SF, as there are provocative connections to our current debates about science, gender and nature. Reading both *Frankenstein* and the author's account of the circumstances of its production is a significant experience for any feminist, not least one interested in science. In brief, Mary, together with her husband the poet Percy Shelley, was in Switzerland as part of a house party including Byron. At the latter's suggestion each agreed to write a ghost story. Mary, at the time only nineteen, described how she spent the evenings listening to the philosophical debates between the men, including the possibilities of life being generated through electricity. She was also reading the current scientific books including that of Humphrey Davy. She describes how she suddenly realised that this Promethean theme provided the material and the inspiration to write her ghost story. In considerable measure she was politically and philosophically her parents' child, for her book subtly fuses Mary Wollstonecraft's feminism and William Godwin's theory of the necessity of social benevolence.

The Monster is unquestionably the product of Frankenstein's science. But it is Shelly's explanation of why the Monster not only has to look, but be, monstrous which speaks of her proto-feminism. For she is unequivocal that it is Frankenstein's denial of love to his creation which leads to the Monster's evil practices. Indeed the Monster in a dialogue with his creator explicitly charges him with rejection. Murderous science is the inexorable outcome when the scientist has no nurturant love, no moral responsibility for his creation. Painfully Shelly's Monster insists he could have been different had Frankenstein loved him.

Her Gothic story today, two centuries on, still creates an indelibly ambiguous and disturbing elsewhere, like and unlike daily life. What then seemed to be a wild speculation on the life generating powers of electricity is routinely echoed in today's speculative proposals by leading molecular biologists heading up major biotechnology firms. Thus, leading US molecular biologist Craig Venter of Celera recently suggested that it would be possible to construct artificial life through bio-engineering. Such a Promethean proposal is likely to arouse technical even ethical criticism from his fellow biologists, intense moral concern by the politicians and the media, and the bioethical industry will go into overdrive, but there is a sense in which the public feels uncomfortably aware that it is only time before the impossible arrives. But Mary Shelley took us there in the imagination two centuries ago and her question of who is to responsibly love and care for these new life forms, these new Monsters, remains.

The Happy Marriage of Science and Socialism

It has become something of a truism that we live in a risk society where risk itself is produced by scientific and technological development; where disasters and the threat of disaster have made us highly conscious of the many threats to the environment. Global warming, nuclear pollution, genetically modified food and BSE are among the new scientifically generated hazards. Nothing could be in greater contrast than the view of science held by the thirties Marxists. This took for granted, and elaborated unflinchingly, the modernist project of the domination of Nature. Thus while writing futurist accounts of science was seen by many leftists as a both serious and pleasurable way of communicating technological possibilities, there was no reflection in these visions of what such transformations might mean negatively for either green nature or sometimes of even our own bodily natures. The futurist vision of the Marxist and crystallographer Desmond Bernal - a figure incidentally who can be said with some justice to have invented the entire field of the social studies of science - is exemplary. Bernal's thesis, spelt out in *The Social Function of Science* (1939) is that scientific rationality and socialism are part of an inexorable happy and progressive marriage. Such a sanguine view of the science and social relations, whether socialism as a longed for political objective or what was so ironically called 'actually existing socialism', sounds remote and unconvincing to a contemporary audience so accustomed to immense and increasing ecological risk from science and technology.

Nonetheless in the thirties this vision of the happy marriage of science and socialism inspired a generation of socially conscious scientists, for whom socialism and science went hand in hand. For this generation more, not less, science and scientific rationality was what they wanted. Thus they opposed Nazi anti-Semitism, not directly on the grounds of its racism, but on the grounds of its poisonous irrationality. Reason belonged to the left, for these were the true heirs of the Enlightenment. For twentieth century people the Nazi death camps were the watershed. These cut off reason and rationality from being part of the language of social progress. For here the machinery of killing millions of scientifically designated *Untermenschen* was organised with intense and inhuman rationality.

But this appalling history was yet to take place and Bernal's essay *The World the Flesh and the Devil* (1929) was written at the height of confidence in the enlightenment. He sees the future where humanity in all its messy diversity has been transformed into one vast disembodied intelligence. Ironically Bernal, who was himself to experience strokes in his upper middle age and was to become an intelligence trapped within a body which could no longer respond, wrote as a young man with enthusiasm of a future aetherialised consciousness:

> The new life would be more plastic, more controllable, and at the same time more variable and more permanent than that produced by the triumphant opportunism of nature. Bit by bit the heritage in the direct line of mankind - the heritage of the original life emerging on the face of the world - would dwindle, and in the end disappear effectively, being preserved perhaps as some curious relic, while the new life which

conserves none of the substance and all of the spirit of the old would take its place and continue its development...Consciousness itself may end or vanish in a humanity that has become completely aetherialised, losing the close-knit organism, becoming masses of atoms in space communicating by radiation and ultimately resolving itself entirely into light (Bernal, 1929 p.46).

The biblical claim of light as alpha and the beginning, is transformed by Bernal into light as omega and culmination. Intelligence is dematerialised, the goal of disembodiment has arrived. It is as if today's artificial intelligencers have entirely won through, mere messy bodily nature can be discarded as unnecessary. Bernal's near mystical vision of the intellect as light is not easy to locate within a materialist tradition, consequently historical materialists tend to down play *The World, the Flesh and the Devil* on the grounds that he was only half way to embracing Marxism and this was still the work of an idealist. This is too easy an escape, rather Bernal's mysticism should be read as part of the physicists' long preoccupation with God and the cosmos. It is not by chance that Bernal was a crystallographer, a field sometimes also spoken of as physical chemistry, certainly one a great deal closer to physics than to biology. As Margaret Wertheim argues in *Pythagoras' Trousers* (1997) this mystical preoccupation is peculiar to physicists, whether an Isaac Newton in the seventeenth century, a Stephen Hawkins today, or for that matter a Desmond Bernal in the late twenties. As Wertheim tartly notes, physicists in their preoccupation with God have succeeded in appointing themselves as His priests - no small impediment to women wanting to study the physical universe.

Feminists have often observed that men unconscious of the injustices of gender envisage utopias which are all too commonly women's dystopias. Bernal's futurist project goes - as befits the most influential science theorist of his time - much further and deeper than this, for he wants to dissolve intelligence into masculinist abstraction itself. His dream of a universal disembodied intelligence is the precise antithesis of the dream of responsible situated and embodied knowledge held by so many feminists half a century later.

By contrast biologists - ever since Darwin - have tended to be materialist, though not necessarily any more friendly to women even in their futurist visions. Thus because of the advances in embryology in the twenties and thirties, biologists became interested in the possibility of cloning human beings. All too many echoed the androcentric preoccupations of their gender. Like the leading geneticist Herman Müller (1936), they spoke of the possibility of cloning their cultural and political heroes (almost never heroines). Indeed if they had been able to realise their favourite wish there would have been a fearful number of little Lenins conceived in petri dishes. After Lenin, the most preferred clonee for male left and liberal scientists alike was Einstein. This androcentric vision was however not permitted to flow on entirely uninterrupted. Both the journalist and feminist Charlotte Haldane (1927) and also her husband, the geneticist and Marxist, J.B.S. Haldane (1924) wrote about the future of genetic engineering. They saw the latter as an entirely feasible development arising from contemporary genetics. The ultimate mysteries of light were not for them. She concentrated on the negative implications of sex selection for women while he celebrated the possibilities of cloning clever men and (uniquely among male scientific futurists leftist or otherwise) clever women.

While the thirties Marxists were prescient in forecasting future scientific and technological developments so that artificial intelligence and genomics (to say nothing of cloning) are part of their crystal ball gazing, what none of them do is even hint at an alternative relation to nature. In the longer story this is slightly unfair to Haldane, for after he left the Communist Party over the Lysenko affair, he went to work in India. There he became interested in developing a non-violent biology, a preoccupation much closer to the project of today's environmentalist movement of living in a non-dominating relationship to nature. But this non-violent biology was a post-war development. In the thirties, Haldane, like the other Marxist futurist writers simply took it for granted that nature is there to be dominated by Man. Nature will inexorably become more and more social and that this, in socialist hands, is an unqualifiedly good.

The Haldanes' move to introduce gender into futurist visions around human reproduction was reinforced by the bleak dystopic SF novel of Katherine Burdekin, *Swastika Night* (1937). Here Burdekin envisages a future world post-Nazi conquest, and points to the entirely dystopic future for women by the extreme cult of masculinity she saw as the core of Nazism. For that matter Dora Russell's *Hypatia or Woman and Knowledge* (1925) offered a feminist utopia of a non-authoritarian and non-technocratic society, sharply at odds with the prevailing masculinist and technocratic enthusiasms of the Marxists. Feminists were complaining (as usual), and assiduously trying to weaken the belief in the happy marriage of science and socialism so enthusiastically promulgated by the Marxist visionaries.

Le Guin and Humanist Science Fiction

By the mid-twentieth century the production of SF had for the most part become both a specialist literary activity and also both andro-centric and andro-authored. There were brilliant exceptions like the pseudonymous 'James Tiptree' and Ursula Le Guin. Unlike mainstream SF, where the excited focus was on technological innovation and where the social is treated as merely intensified present social relations - just more hierarchical, more racist and more sexist - alternative SF such as that of Le Guin (1974; 1979; 1989; 1992) opens up new possibilities facilitated by, but not limited to, technological change. She, for example, envisaged a democratic anarchistic utopia where gender is softened to androgyny and reproduction is desexualised. In her critical writing Le Guin has denounced the mainstream's obsession with technological change and relentless social conservativism trenchantly:

> The only social change presented by most SF has been towards authoritarianism, the domination of the ignorant masses by a powerful elite - sometimes presented as a warning, but often complacently...in general American SF has assumed a permanent hierarchy of superiors and inferiors, with rich, ambitious, aggressive males at the top, then a great gap, and then at the bottom the poor, the undereducated, the faceless masses and all the women (LeGuin, 1979 p.99).

The pervading quality was a macho enthusiasm for the technology of domination and its equation of technological advance with progress. Generally politically reactionary, it was frequently racist and almost invariably sexist. The dialogue between science and science fiction has produced two very different traditions in main/male stream SF. One is triumphalist and technicist, a socially unreflecting mediation; the other is preoccupied with the social implications of impending scientific and technological change and takes the form of utopian fictions/essays or dystopic warnings. Technicist science fiction can manage perfectly well without a utopian thought in its head, acting instead simply as a magnifying mirror to existing society, its boyish enthusiasm for toys all too evident.

The agenda of a political SF, like that of the humanist Le Guin, was concerned with both technological and also social transformation. Despite the publication of Rachel Carson's *Silent Spring* (1962) it was some time before ecological considerations entered either SF or for that matter science studies. Nature was the absent dimension - a taken for granted back drop which although constantly transformed by culture, would some how always be there, a repository of inexhaustible wealth and self-recovery.

During the thirties and beyond, only the Frankfurt school continued to question scientific rationality and social progress as part of modernity's happy marriage. It was not just the scientists for whom it could be argued that this was a self serving ideology, but the philosophers, sociologists and historians of science, indeed most social thinkers who saw science (especially basic or 'pure' science) as ethically neutral. What mattered, they argued, was its social use or abuse. This conception of science as neutral was only challenged by the intense questioning of what was seen as the genocidic science and technology deployed in the Vietnam war. This period opened an entire fresh wave of criticism from the new left and from the burgeoning women's movement. Although the radical science movement interrogated the neutrality of science, asking just how far was science neutral, objective, independent from social and cultural values, it was the writers within the women's liberation movement who so brilliantly developed a distinctively feminist SF which raised all these theoretical debates within an extraordinarily accessible medium.

The Birth of Feminist Science Fiction

Joanna Russ, herself one of the most significant of second wave feminist SF writers, very early on identified SF as a genre peculiarly appropriate for feminists. She argued that it provided a vehicle for exploring the pressing anxieties and experiences of feminists concerning science and technology in a way which was not possible within more traditional genres, such as the novel, where the form constricted imagination (Russ, 1974).

Russ was right, the new FSF was central within the torrent of creative imagination unleashed by the women's liberation movement. Two names stand out in this early period as powerful writers and as representing the feminisms of the seventies. These two were the socialist feminist and realist Marge Piercy and the radical feminist and postmodernist Joanna Russ. Seventies feminists had no truck with fixed blue-prints for the future, their worlds were radically indeterminate and always to be fought for. Thus both Piercy's *Woman on the Edge of Time* (1979) and Russ's *The Female Man* (1975) depict simultaneous androcentric dystopias and feminist utopias in opposition. Piercy's heroine Connie bears the multilayered oppressions of race, class and gender. Piercy sets about envisaging an imaginary utopia in which reproductive technology and other biological interventions could be used to socialise pregnancy and motherhood.

Feminist dreams and biomedical realities met in 1978 in all their ambiguity with the birth of the first IVF baby Louise Brown.

As a socialist feminist Piercy is sensitive to issues of class; her involvement with the US radical science movement, and her engagement in the struggle against scientific racism, added anti-racism to her political agenda long before white feminism began to give up its taken for granted eurocentricity, and gave a sophistication to her analysis of science and technology lacking in authors such as Firestone (1971). In her utopia Mattapoisett, the link between biological motherhood and the societal need for new people is broken by turning, just as Firestone does, to extrauterine conception. The difference is that where Firestone saw technology as outside society and as an autonomous and neutral agent of change, Piercy locates the technology as an integral part of the new classless and raceless society organized around a post-gendered system. Her direct links with the civil rights movement and the radical science movement inform her text.

In Mattapoisett she plays with genetic engineering in a way that outside of a utopia most of us would find hard to take, remembering the long, brutal and far from complete history of eugenics which even today confronts us in new consumerist mode. She accepts, for example, that human beings take pleasure and comfort in seeing the features of someone they love reproduced in a new being, and has no qualms about proposing that the replacement for the killed Jackrabbit bears some of the beloved's genes. Among contemporary feminists working on genetics, where strong social constructionism threatens to squeeze out a recognisable sense of embodiment, only Barbara Katz Rothman (1999) reflects on our recognition of faces and bodies with such directness.

Piercy gives to Luciente a figure from her utopia a vision of the choices:

> It's that race between technology in the service of those who control, and insurgency - those who want to change the society in our direction. In your time the physical sciences had delivered the weapons technology. But the crux we think, is in the biological sciences. Control of genetics. Technology of brain control. Birth to death surveillance. Chemical control through psychoactive drugs and neurotransmitters (Piercy, 1979 p.229).

Twenty five years on Peircy's predictions look to be on target except that the proliferation of technological risk, not least in bioengineering, has also laid massive siege to green Nature. She understands that not only is nature modified continuously by culture - and that includes our own nature - but that our conceptions of what is natural and what is cultural themselves undergo constant negotiation. Her SF novel thus anticipates and confirms the point made and remade by the feminist critique of science over the last two decades. Her utopia is ecologically sensitive and people live in harmony with their environment. They also develop new relations with animals, or rather some animals (conversations between cats and people are possible, whereas a dinner of roast goose remains in prevegetarian carnivorous glory). Sexuality is polymorphously permissive, where desire includes love, liking, pleasure and comfort.

By contrast, Russ's (1975) radical feminist pursuit of a non-patriarchal society, is centrally preoccupied with sexuality and gender, while race and class and nature get scant attention. She is less optimistic than Piercy about the chance of a new society with a reconstructed sex-gender - let alone post-gendered - system. Her utopia is called Whileaway, because the men are 'away', having died through a sex linked genetic disorder. It is thus necessarily and pleasurably separatist. Men, writes Russ, 'hog the good things of this world'. But where Piercy writes as a realist, Russ is a post-modernist, not simply acknowledging fractured identity but deliberately splitting herself into four - whose names all begin with J - to play out four simultaneous worlds, yet with the Js meeting and travelling together through the hostile terrain of Manland. By contrast with Piercy who works through realism and identification with a heroine who bears the multilayered oppressions of America, Russ, except for brief moments (like the love making between Janet and Laura which is privileged both politically and in literary style) precludes identification. Instead the reader is constantly distanced in an explosion of witty and savage writing.

Her title *The Female Man* (1975) evokes all those women writers of SF from C.L. Moore to James Tiptree who wrote as female men. Two of the Js are drawn from the present or near present - Joanna (the writer) and Jeannine, whose total existence revolves round 'the Man'; Janet comes from Whileaway, and Jael is a warrior. In Manland where there are no actual women but only modified men, the confrontation between Boss-man and Jael sends up an all too familiar story. Boss-man, over-excited by a real woman, talks first of equality, then fucking, then proposes to rape her. As he demonstrably has failed to take no for an answer, Jael kills him. When one of the other Js, asks if there wasn't an alternative, Jael replies that she enjoyed killing him. With one move Russ decisively breaks the link between the male sex and violence. To a feminist the scene is both comic and emancipatory; a humanist has a harder time.

Russ eschews that restricted variant of feminist science fiction where the utopia merely ensures role reversal; instead she burlesques it by creating a scene between Jael and her pretty boy live-in lover only to reveal that Dave is in fact a robot. Janet's comment is left to sum up: 'Good Lord', she says

'Is that all?'. Compulsory heterosexuality is integral to Jeannine's oppression and as a mere robotic substitute within Jael's role-reversed world. Within Russ's utopia, the polymorphous permissiveness of Piercy is insufficient to deal with the problem of men. Only lesbianism offers the possibility of fusing mutual caring and passion.

Although Sally Gearhart's SF writing is criticised (Paulser, 1984), not without reason, as that of a propagandist using SF as her vehicle, her voice is precisely that of the eighties turn to new age ecofeminism. As such it is the antithesis of thirties Marxism. There is no happy marriage, in or out of any transformed social relations. Holding high-tech reproduction in distaste, her utopian inhabitants, the hill women, reproduce by gene-merging, a distinctly mystical process despite its scientised language. Her ecofeminism goes far beyond the relatively modest project of Piercy. Where Piercy wants green trees and clean rivers, Gearhart's Hillwomen who have a collective rather than individual identity, communicate directly with the sky and trees. Indeed the sense of where one species ends and another begins is blurred, a blurring that has developed politically with the rise of the animal liberation movement. Gearhart is concerned not simply to establish a respectful relationship to the environment, which is a common enough aspect of FSF, but wants to establish nothing less than a common means of communication between nature and culture to overcome the division between them. Her ecofeminism can only be sustained by the new natural/cultural system of the Wanderground.

With the single and exceptional voice of Marge Piercy, these white feminists do not go beyond a non-racist stance. Even the intensely political Gearhart, cannot take anti-racism on board. She praises Piercy's work and goes on to reflect on her own:

> For me moving myself out of a non-racist stance and into an anti-racist one is like trying to push an idle steam roller. I can't get moving, it seems hopeless, and it's easier to do something that I have more passion about, more success in doing (Gerhart, 1984, p.301).

Indeed that the two women of colour who appear in the Wanderground are rescued by white Hillwomen is not in itself an overly encouraging signal to black women. This may be the price of a concept of sisterhood that claims a solidarity which, beginning by de-emphasizing individuality, ends by meeting racism with a replay of the white woman's burden. A utopia created from within a eurocentric feminist perspectives, in which the distinctive experiences of women of colour and black women are erased -

can only be dystopic. A feminist eurocentric utopia becomes a black women's dystopia.

It is for this reason that I want to turn to Octavia Butler as my last SF writer. Her books, beginning in 1980 with *Wild Seed*, saw a strongly African - American perspective enter the feminist SF world. She takes themes of slavery, time travelling, taking over bodies, and survival in a post-nuclear and post-colonial world where the boundaries between people, animals, machines, and even beings from outer space are permeable.

In Dawn (1987) the second in her *Trilogy Xenogenesis*, the central figure is Lilith (not by chance named for the traditional conflation of the Hebrew demon who destroys children and Adam's first wife created separately from him and who refused to have sex missionary style). She is rescued from the aftermath of nuclear holocaust by extraterrestrial beings, the Oankali. This young African American woman, already a widow through an accident in which her son was also killed, becomes the non-consenting reproductive partner of the Oankali. While the extraterrestrials try to repair Earth, the human beings are stored in a state of hibernation on the Oankali space ship - itself a living structure. They are awakened in small groups to learn how to deal with their new lives on the ship. These survivors, including Lilith, find themselves faced with the choice of either privileging their humanity and becoming Resisters who refuse to collaborate with the Oankali - effectively a policy of suicide - or of entering into some sort of connection in which human autonomy is threatened to the point of extinction.

Careful to make a number of the Resisters sympathetic figures, though showing the deadly nature of their refusal as their genetic inheritance commits them to hierarchical behaviour (in this showing herself an unequivocal biological determinist), Butler carefully pulls the reader's sympathies towards the far from free exchange that Lilith enters with the Oankali. The exchange reaches new and disturbing levels of intimacy, for the Oankali are gene-traders. They constantly sense, modify and exchange genes. Initially horrified by humans as inherently drawn to lethal violence, the co-operative Oankali become fascinated by humans' genetic complexity and danger. Where cancer is a symbol of terror for the humans, to the Oankali, cancer-bearing cells such as those of Lilith, are a resource for technologies of metamorphosis and regeneration.

Butler's imagined world is postmodernist in its recognition that we are compelled to live with profound ethical and cultural ambiguities. Thus Butler lets the reader share both Lilith's revulsion at being pregnant with a non human child and her learning an unpossessive affection for her strange hybrid offspring. Her son Akin gradually realises the changes taking place within the Oankalis are chosen; those of the humans are unfree. Genetic determinism only operates for humans. As he puts it 'we', and by that 'we' he means that part of him which is Oankali, 'consume' human beings, they, the humans, do not consume us. Yet Akin as part human, also understands the desire for autonomy by the Resisters and the anger they feel against the Oankali for denying them.

Where Mary Shelly's Monster was a creation from a feminist reading of early nineteenth century science in which electricity generated life, Butler's Lilith and her son Akin are located in a feminist and postcolonial reading of the embryology and genetics of the late twentieth century set in a post-holocaust world. Yet where Shelly's patriarchal science excluded love, thus creating a fallen angel who became a devil, an ultimately dystopic vision, Butler's alternative Oankali science creates hybrid beings who do have some sort of feeling relationship with others - and begin to pick their way. While it would be too strong to say Butler's space ship is a utopia, it is not without hope.

Her reading of the future requires that the old lines between nature and culture, between species, and between structures and living things have to change. Social practices - racism, as much as sexuality and reproduction - which were part of the past, become irrelevant. Butler opens a kind of Pandora's box where she shows us that the worst things (i.e. global nuclear holocaust) contain some room for hope. Lilith, and even more her construct son, Akin, have to negotiate both the consequences and the hope.

In the third volume of the trilogy, *Adulthood Rites* (1988) Butler seems to resist claiming that the future can only be hybrid. Unlike Donna Haraway's (1985) celebration of cyborgs which unhesitatingly embraces the taking down of the boundaries between people, animals and machines, Butler has hesitations and leaves open the possibility that autonomous human beings can create a future. Thus although because of his special gifts as a hybrid, Akin is able to see and respect the longing of the Resisters, he has himself has to become more Oankali-like in order to help them. While some readers, notably Haraway see Butler as sharing her enthusiasm for pregnancy with another species, I read her more ambiguously as suggesting that hybrids may indeed be the future but that she also understands the longing for species identity and the ongoing dream of a human future. Butler leaves Akin to negotiate this still undecided future.

Sustaining or Losing our Memory of the Future?

By contrast with the masculinist utopias and science fiction, these feminist visions of transformed natural and social worlds emphasise again and again that a particular form technoscience is not inevitable; as part of culture it is socially constructed and as such, is continuously subject to reconstruction. Thus however powerful the present turn towards the new technosciences of informatics and biotechnology, these are more socially malleable than their protagonists claim. It was above all questions of reproduction, both human and global, that were addressed head on by the golden age of FSF. These extraordinarily diverse scenarios in all their radical indeterminism nourished feminists' capacity to rehearse strategies against hegemony. Yet the move by biology into the industrial mode of production of Big Science has dramatically changed the landscape. Biotechnology - DNA genetics - has huge implications for reproduction and the financial and scientific stakes are much bigger than ever before.

Sadly by the nineties this spring of this imaginative combatative FSF writing, exploring the pressing anxieties and experiences of feminists concerning science and technology, had weakened in its flow. Post-feminism and the feminist backlash diminished its audience, even the plurality of feminist bookshops shrank to a few stalwart survivors. Even before we reached the millennium some of the most widely seen and read science futurist accounts were already coming from the right, not the left and not from feminism. The film *Gattaca* showed the Hollywood version of the genetically modified society, with an unmodified man as hero to fight his John Wayne-like way through to the top and get the girl. Nonetheless the take home message of this ill written and performed film is unambiguous: the human future is genetically modified. For that matter *Remaking Eden* (1998) written by US molecular biologist Lee Silver, offers a remarkably similar picture as commodified eugenics in the context of a highly marketised society mean that pressurise parents choose to enhance their children's genes. Middle class parents buy social privilege through fee paying schools, what is so very different, Silver questions, about buying genetic privilege? These inexorable and dystopic visions of the future intensify an increasingly geneticised culture. Yet this genetised culture is scarcely getting everything its own way, for there is also huge and global resistance above all to the genetic manipulation of green nature but also to human biopiracy. More academic critique of biotechnology and the geneticisation of culture as well as more protest and ferment may provide the preconditions for our most imaginative writers to return to SF to offer us new and compelling alternative visions. In hard times we cannot afford to lose our memory of the future.

Note

1. This chapter reworks some of the themes originally developed in Hilary Rose's book *Love, Power, and Knowledge*, Polity 1994.

References

Bernal, J.D. (1929) *The World the Flesh and the Devil*, Cape, London.

Bernal, J.D. (1939) *The Social Function of Science*, Routledge and Kegan Paul, London.

Burdekin, K. (1937) *Swastika Night*, Lawrence and Wishart, London.

Butler, O. (1980) *Wild Seed: Xenogenesis I*, Doubleday, New York.

Butler, O. (1987) *Dawn: Xenogenesis II*, Warner, New York.

Butler, O. (1988) *Adulthood Rites: Xenogenesis III*, Warner, New York.

Carson, R. (1962) *Silent Spring*, Houghton Mifflin, Boston, Mass.

Firestone, S. (1971) *The Dialectic of Sex: The Case for Feminist Revolution*, Cape, London.

Gearhart, S. (1985) *The Wanderground*, The Women's Press, London.

Gearhart, S. (1984) Future Visions Today's Politics: Feminist Utopias in Review, in Rorlich, R. and Hoffman, E.H. (eds) *Women in Search of Utopia*, Schocken, New York.

Haldane, C. (1927) *Man's World,* Doran, New York.

Haldane, J.B.S. (1924) *Daedalus, or Science and the Future*, Cape, London.

Haraway, D. (1985) A manifesto for cyborgs: Science technology and socialist feminism in the 1980s, *Socialist Review*, 80, 65-107.

Irwin, R. (2000) Science fiction Studies *Times Literary Supplement* 5056, February 25th, 30.

Katz Rothman, B. (1998) *Genetic Maps and Human Imaginations*, Norton, New York.

Le Guin, U. (1974) *The Dispossessed*, Gollancz, London.

Le Guin, U. (1979) *The Language of the Night*, Perigree, New York.

Le Guin, U. (1989) *Dancing at the Edge of the World*, Gollancz, London.

Le Guin, U. (1992) *The Left Hand of Darkness*, Orbit, London.

Lessing, D. (1979) *Colonised Planet 5 Shikasta*, Cape, London.

Lessing, D. (1980) *The Marriages between Zones Three, Four and Five*, Cape, London.

Lessing, D. (1981) *The Sirian Experiments: The report by Ambien II, of the Five*, Cape, London.

Müller, H.J. (1936) *Out of the Night: A Biologist's View of the Future*, Gollancz, London.

Paulser, I.L. (1984) Can women fly? Vonda McIntyre's *Dreamsnake* and Sally Gearhart's *The Wanderground, Women's Studies International Forum* , 7, 2, 103-110.

Piercy, M. (1979) *Woman On the Edge of Time*, The Women's Press, London.

Rose, H. (1994) *Love, Power and Knowledge. Towards a Feminist Transformation of the Sciences*, Polity Press, Cambridge.

Russ, J. (1974) What can a heroine do? Or, why women can't write, in Glasgow, J. and Ingram A. (eds), *Courage and Tools: The Florence Howe Award for Feminist Scholarship 1974-1989*, Modern Language Association of America, New York.

Russ, J. (1985) *The Female Man*, The Women's Press, London.

Russell, D. (1925) *Hypatia, or Woman and Knowledge*, Cape, London.

Shelley, M. (1819, [1969]) *Frankenstein: or The Modern Prometheus*, Oxford University Press, Oxford.

Silver, L. (1998) *Remaking Eden*, Weisenfled and Nicolson, London.

Wertheim (1997) *Pythaoras' Trousers*, Forth Estate, London.

Wolf, C. (1977) *The Reader and the Writer*, Seven Seas, Berlin.

Part Four

Future Science, Future Policy and the Management of Uncertainty

9 Scripts for the Future: Using Innovation Studies to Design Foresight Tools

BASTIAAN DE LAAT

Foresight has increasingly become an important support for decision making on, and management of, technology. Foresight - and earlier 'forecasting' - is often performed by the aid of specific methodologies which aim to predict the future or sketch alternative futures. Delphi studies, scenario writing or regression analysis tries to better the capture the elements which are expected to become important in tomorrow's world and the conditions under which they will become so. Following Van Lente (1993), foresight methods therewith help to create expectations on the future evolution of technologies and express a promise on their future potential.

Yet whatever their method or approach, the 'promises' traditional future studies come up with often only give us *half* of the story on the future. They effectively tell us which macro-evolutions (as in economic forecasting) or scientific breakthroughs (as in Delphi studies) we can expect. They may also sketch boundary conditions that future technological evolutions will - or should - be subject to. In energy technology modelling, for instance, such boundary conditions concern issues such as maximum CO_2 emissions, world population, per capita energy consumption, or more recently the effects of liberalisation on energy markets. Traditional foresight may even assess undesired side effects of technologies and warn about the ethical problems involved with gene therapies or the Information Society. Traditional foresight methods however have great difficulties telling us simply what the world will look like if a policy maker, a researcher or an engineer, decides to promote this or that technological choice. Until recently, futurologists '...have shown plodding pedantry, in a preoccupation with dredging up definitions. Dodging real-life concerns, they have been engrossed in pencil-and-paper games, the intellectual doodling dignified by names from the Greek ...'.[1] For real images of the future one had better buy books by Verne, Orwell, or Crighton.

The results of 25 years of innovation studies appear to be of a great help to think about technology foresight in a manner complementary to traditional futures thinking. The basic idea that will be put forward in this chapter is easy to grasp – and eventually starts to be more and more and more adopted in policy thinking: every technical choice implicitly involves a hypothesis on how, *sociotechnically*, the future may be organised. For instance, funding research on a battery for an electric vehicle is not only a technical matter. Such a choice implies that at the pump, electricity will replace gas, that the battery will replace the car's tank, and that the driver is patient enough to wait for a (long, long) refill at an 'electricity'-station. For an ecologist, an electric vehicle probably fulfils several elements of his or her normative future scenario: a car that is less polluting, less noisy. However, a technical object goes beyond such promises as it contains a number of hypotheses on the socio-technical environment it has to be inserted in, and which are not given by traditional tools and methods. This set of hypotheses can be called a future, and therefore as yet 'fictive' script. Over the past few years, I have used the idea of fictive scripts to analyse policy settings (e.g. for national agencies to compare research programmes). But it has also proven an interesting concept for educational purposes, especially to make future policy makers and research managers aware of some of the implications associated with their work. This chapter gives some theoretical background, and briefly discusses some of the experiences with this new way of really looking 'into' (and not 'at'...) different potential future worlds.

Foresight and Strategic Analysis in Constructionist Perspective

An important feature of innovation studies developed over the last decade is that they take a so-called constructionist (or constructivist) stance.[2] Contrary to what is often suggested, constructionism does not equal an unbounded relativism.[3] 'Socially' constructed facts may become hard facts in the end – even if they stem from a 'soft' scientific practice like strategic analysis or foresight.

However, adopting a constructionist view on futures thinking puts traditional ways of futures analysis into new light. The general argument is not entirely new. A comparable view was exposed in a lengthy and critical essay by Ogilvy (1996). This author proposes that future studies should no longer be based on the 'firm foundations of accepted science' (1992 p.6), or what he calls the 'positivistic programme' in future studies. This programme is characterised by the efforts of those seeking to identify general laws which would permit prediction and the deployment of a 'methodology which gravitates toward better measurement techniques, improved polling, or statistical techniques and modelling tools that might uncover law-like regularities amidst masses of data' (1996 p.11). Such efforts, the author suggests, '...are fundamentally misguided as putative answers to the current questions about futures methodology' (p.11). The alternative proposed by Ogilvy is to base future studies on anthropological principles,[4] that is, they should not impose any *a priori* method upon those participating in, or being the subject of, a future study - and thus neither a scientific method. Leaving aside whether the solution he adopts in the end – a normative scenario in the form of a narrative - really fulfils his anti-positivistic programme, I do subscribe to Ogilvy's main argument. That is, adopting a constructionist view on strategic analysis and foresight, based upon considerations similar to the ones mentioned in Ogilvy's essay, implies that futures methodology can no longer rely a priori on the positivistic-scientistic formalisms (whether they are economic or technical) which are normally used for this purpose. Just like the sociologist of science who will not take science as *explanans* but as *explanandum*, a futurologist could try to understand the construction of meaning through futures methods instead of taking for granted their results.

The following section works out this argument from a theoretical perspective. This will be done along two lines. The first focuses upon the *prescriptive literature* ('which methods are proposed in order to carry out strategic analysis') the second on the *descriptive literature* ('how do actors use these methods in practice').

Two Ways of Viewing Methods for Strategic Analysis and Foresight

Two separate bodies of literature relating to the problem of strategic analysis of (technological) options can be distinguished. The first, a substantive field, concerns the literature on *methods* used for ranking (research) priorities. This is the literature that prescribes the methods to be used within the process of strategic evaluation and foresight. The second body of literature, much smaller, takes more of a 'meta'-position and analyses the role of such methods *within* processes of strategic evaluation and foresight, or more generally, within innovation processes.

Traditional Methods Concentrate on 'Stabilised Settings'

A vast majority (not to say most) of the formal methods for future studies, forecasting, or foresight, which today are still used, originate in the 1950s and 60s, when they were developed *for military purposes*, by the Rand cooperation. Most of these methods such as Delphi and Scenario analysis, and the less well known Cross-impact Analysis or Relevance Tree Analysis are still in use.[5] A detailed analysis of the literature[6] shows a shift away from foresight to a more retrospective interest in the development of technical innovations during the 1980s - even in typical foresight journals like *TFSC*, *Futures*, *Research Policy* and *TASM*. This shift in interest may be explained by a certain loss of faith in the development of formal models and methodologies amongst foresight scholars (see for instance Coates, 1989; Gordon, 1989). Another reason could be that conventional models had in the meantime been adopted by actors other than those belonging to the academic community (firms for instance) leading thus to lower visibility in the academic literature. A third potential reason for this loss in interest could be the turn toward more interactive approaches and the increased focus on 'group decision making techniques' (Crookall, 1995) as well as to more integrated 'technology assessment' approaches (Smits and Leyten, 1991). At the turn of the 1990s however, a renewed interest in technology foresight can be observed amidst both scholars as well as policy makers.

Futures methods are difficult to categorise and, in studying the literature, it turns out that different authors classify similar methods differently. In fact only the meaning of 'forecasting' is relatively well shared by different scholars in the field, most agreeing that it relates to the *quantified prediction* of future events, arrived at through formal modelling and/or expert opinion. But even then boundaries differ from author to author, and moreover, meanings seem to shift over time. For example, the often-cited OECD publication by Jantsch (1967) distinguishes between intuitive reflection *vs.* exploratory forecasting *vs.* normative forecasting. For Godet (1985) however, the scenario-method is central, scenario-construction depending on the results of a broad panorama of other methods and information gathering strategies. Amara (1989) distinguishes between expert judgement, structural modelling and descriptive forecasting. But in practice the three are not positioned on the same level since expert judgement very often appears as input to structural modelling and descriptive forecasting.[7] As an alternative, others propose distinctions between forecasting and the actual planning process, between prévision/fore*casting* and prospective/fore*sight*, or again between strategic analysis and foresight, by saying that the former is a central element of the latter (see respectively Godet, 1985; Martin, 1995). In any case, the analysis of the literature makes clear that terms are not so precise and that, depending on the situation, methods which should clearly be separated in the eyes of one author intermingle and interact for another. For instance, scenarios sometimes integrate quantitative 'forecasts', sometimes they are simply written by a 'bunch of clever people', and in several cases they are not an end result, as one would expect, but in serve as starting point, as input to *other* methods. Similarly, whereas the label 'foresight' in the 1990s was becoming more and more exclusively reserved for designating foresight on *national* levels (Georghiou, 1995; Cameron et al., 1996), exceptions to this rule can easily be found.[8] Also for the common denominator 'strategic analysis', things appear not to be so commonly shared at all.[9] So, whereas in local situations distinctions may be made, no general, 'theoretical' demarcations can be drawn. Formal definitions may exist, but in practice different labels are used for a variety of foresight activities, in an even greater variety of combinations.

Two further conclusions can be drawn from the analysis of the prescriptive literature. As confirmed time and again by handbooks (e.g. Porter et al., 1991, or Twiss, 1994) or review articles (Coates et al., 1994; Drejer 1996) futures methods apply to the *external environment* of the 'client' of the exercise - a company, a public authority, a research programme, or other type of organisation. Forrest (1991) best shows this feature. In a review of different (firm-level) innovation models, this author shows that strategic analysis often constitutes a formal nexus between the organisation and its (future) external environment. This future environment cannot be modified by the firm and hence is to be simply 'reacted upon'. However, anticipations are not mere reactions to representations of an outside world. An actor like an industrial firm or a research institute is not passive. A complementary element to strategic analysis can therefore be identified. This is to look *inside* the organisation for which a strategic analysis is carried out. Though practised since a long time by management consultants working for firms (e.g. Hamel and Prahalad, 1994) this domain was until now largely ignored by strategic technology analysts. For the most part, strategic analysts consider it to be the firm's job to develop an appropriate technological response to the results of a future study. The concrete construction of technologies is therewith delegated to practical technology management - and, correspondingly, to a separate set of academic journals. It has thus been eliminated from the scope of the methodologist in futures studies - the two elements (future studies and technological choice) are relatively disconnected (Amara, 1989). In this regard, a suggestion made in 1986 by the American futures scholar Bright has passed relatively unnoticed. Bright proposed that anticipations of external environments are made *within* research and within the organisations which put research into practice (1986). Following up on this idea, analysis should not only focus upon the future evolution of external environments, but, symmetrically, be 'endogenous' as well. This idea will be used when discussing innovation studies below.

A second conclusion that can be drawn is that traditional methodology typically conceives of these external environments as being either 'markets' or 'systems'. As will become clear later on, this *de facto* means that they apply to 'stable settings'. And if they do not, they mostly relates to issues very 'far' into the future, as in the case of setting national research priorities, long range planning and so forth. The issues of 'stability' (and irreversibility) will be discussed more precisely below. For now it is enough to note that methods that focus on calculus, systems analysis, or any other parameterisation, work only in settings in which they *can* work, i.e. in situations which for the greater part have been constructed and stabilised. For such situations - for instance predicting sales of products within existing markets following an S-curve - evolutions may be reasonably foreseen. Papers proposing such methods are numerous, and Steele's (1995) is a typical example. Steele's method is able to help decide which products a firm should promote in order to gain the best possible comparative advantage. However, the method can only determine 'the relationship between financial profits and the fundamental engineering and business variables', when the 'unit production cost and unit salesprice are defined' (p.242). In some stages of technological innovation, such data will certainly be at hand. In other cases not. Not only might the data be missing, but there may not even be a consensus about the type of data required for determining the best choice or about the method which should provide such data. In such cases nothing will *a priori* point to one technological route or the other, or justify product A or product B. In determining research priorities for the longer term, policy makers are typically confronted with such situations. In that case, focusing on existing markets or systems dynamics is not of great help.

Methods are Constitutive of Innovation Processes

Except for maybe a handful of forecasters, clinging on to old-fashioned ideals - like those so violently criticised by Irvin (1993) - nobody today believes that nations should determine their research priorities uniquely by applying a formal methodology.[10] As was already shown by the famous *Foresight in Science: Picking the Winners* (Irvine and Martin, 1984), research priority setting on national levels is structured in processes involving different 'stakeholders,' i.e. relevant actors. Only sometimes, and for specific and well-defined purposes, do formal methods intervene in such processes. Even in Japan, for long seen as the 'Mecca' of the Delphi approach by European and American forecasters, formal methods appear to constitute only a small part of future-oriented activity (Bowonder and Miyake, 1993). Successful innovation policy depends more on the formal and informal networks that exist, the interaction between the participants and a continuous thinking over of strategy, than on the perfection of formal methods.

A similar point is made by Georghiou (1996) in discussing the UK Foresight exercise. There the only method used was the Delphi approach, and even then only for a specific part of the process. The greater part of the exercise consisted of iterative and interactive consultative rounds with panels of co-nominated experts and of 'regional workshops.' The same holds for the Netherlands, where on the national level an interactive process (structured discipline-wise) was organised. Formalisation within this process relied upon the scenario approach (OCV, 1996). In sum, at national levels foresight is far from being a single method and such exercises might involve one or several techniques. The choice of the technique does not seem to predetermine the quality of the result. The essence of foresight exercises is that they are collective processes of future thinking toward the definition of priorities. They have not much to do with crystal balls. But they have everything to do with collective construction, as illustrated by the fact that, today, the impact of the UK foresight is evaluated - a fact that besides implicitly acknowledges the 'social' construction of science and technology itself.

Now it is easy to say that formal methods do not have, and maybe even should not have, the last say in such processes. It can be readily observed however that they *are* used, and apparently *do* make a difference. What then is their role? Passing from a national setting to the firm, Galbraith and Merrill (1996) have investigated the use of forecasting techniques and their results within business enterprises as part of strategic analysis and evaluation. The authors have good reasons to take a critical look at futures methodology (1996 p.31):

> the vast majority of empirical work on forecast and modelling dynamics has been limited to comparing the accuracy of different quantitative and qualitative techniques while controlling for the type of data being inputted.

Consequently (1996, p.31) '[i]nternal politics, the "staking out" of positions, corporate culture and ethical attitudes, staff and management training in forecasting techniques... are typically the forgotten dimensions of forecasting and modelling research. Finally, there is an '...increasing difficulty of attempting to reconcile the notion of forecasting and modelling as an objective, scientific, unbiased endeavour with the political realities of modern day management' (1996 p.31).

Galbraith and Merrill's main observation is that forecasts within firms can be subject to manipulation and that 'highly manipulative' forecasting environments are associated with poor training in forecasting techniques (p.33). Moreover, forecasts are typically used to achieve certain goals, like for instance to influence the value of stock price or to get venture capital (p.34). Foresight thus is *performative*. The general, if implicit, point of these authors is that foresight methods are 'socially' produced and used. It is a pity to observe that the authors' retreat, not to intelligent fine-grained analysis, but to a denunciation of the 'wrong' use of forecasting within firms. That is, they suggest that 'bad' forecasting is influenced by 'motives' and 'opportunity', whereas 'good' forecasting follows the rules of scientific method (p.38). They thus fall back on the 'positivistic paradigm' of future studies which, following Ogilvy cited earlier, we precisely are trying to bypass.

A less normative and much more analytical stance is taken by De Man (1987) in his 'attempt at investigating and analysing forecasting [...] from a social science perspective' (p.165). To our knowledge, De Man's study is one of the first attempts to explicitly analyse future studies' methods from a sociological viewpoint. The author in particular studies the role of energy forecasting within policy processes in the Netherlands and the UK from the mid-1960s to the 1980s. He finds that '...the nature of the political contexts is a far more significant explaining factor [for the way in which scenarios are built] than the substance of the models themselves'. If not socially constructed, as the author implicitly suggests,[11] his analysis shows that methods and models are at least socially 'embedded'. However, De Man's picture is not complete: once the scenario, the forecast, or the prediction of future energy consumption, exist, what happens to them...? Although mentioned obliquely,[12] this point is largely ignored by the author - probably because in both the Dutch and British cases he studied the energy scenarios developed were simply not taken up and so remained without any effect. The question therefore is this: suppose that the scenarios *had* been used, how can their role possibly be accounted for? The answer is to be found in Hauptman and Pope (1992) and Van Lente (1993).

Hauptman and Pope (1992) - just like Galbraith and Merrill cited above - are dissatisfied with the way in which forecasting and future studies are conducted:

> most of the best work in technology forecasting research and in forecasting in general is oriented toward developing expert techniques and tools. Consequently, viewed as a specialised discipline and expertise, technology forecasting is quite exclusive. This exclusivity has not facilitated studies of technology forecasting applications and practice by real-life corporate executives. [...] In addition, the focus of attention on exploratory forecasting made the developed knowledge not very useful for normative applications such as technology application and strategy (pp.193-194).

The authors therefore propose to study '...the process of technology anticipation and planning in the context of strategic decision making' (p. 194).

Based on interviews with CEOs from the electromagnetic resonance industry, they conclude that strategic analysis and its methods are completely integrated within processes of decision making and technology planning. Their analysis shows several things. A first conclusion reflects the feeling expressed earlier: CEOs make a clear distinction between the 'external' and 'internal' environment of their organisation – they allow us to draw *boundaries*. Formal tools establish representations of the external environment of firms, thus disconnecting them from anticipations made inside: these remain for the greater part *informal*. Next, the results of futures studies are always only one element amongst many others within technology planning processes. They appear to be hard to separate, even analytically, from the local circumstances from which they emerge. In other words, as we already concluded from De Man, not only a nation's, but also firms' processes to think about the future are *embedded*. A third and related observation is that explicit and implicit – or better, formalised and non-formalised – forecasts are hard to distinguish, not only from the rest of the process, but also from each other: they *intermingle*. Finally, it is observed that the results of this implicit/explicit and embedded planning activity (in the form of heuristics, analogies, scenarios, simulations and so forth), continuously feed into the decision making process (as technological plans, business contingencies, etc.). They are thus *constituent* of such processes.

The observation that future studies and their outcomes are constituent of innovation processes, is very clearly conceptualised in Van Lente (1993). This author studied the role played by promises and expectations in processes of technical innovation. Promises play an important role in agenda building, where they:

> function as a yardstick for the present and as a signpost for the future [...] The implication for the dynamics of concrete developments is that what starts as an **option** can be labelled as a technical **promise**, and may subsequently function as a **requirement** to be achieved, and a **necessity** for technologists to work on, and for others to support. This option-requirement-necessity sequence does not imply that it is an autonomous socio-technical process. The transitions do not occur automatically, but are the result of action and interactions of technologists, firms and governments. The transitions are a consequence of actors assessing what is 'feasible', what is 'obsolete', what is 'necessary', and of their efforts following the assessments. Moreover the transitions are reversible, in principle, and can be made undone - with increasing work and costs, though. The pressure to gain back sunk investments increases also. When after some time much had been invested in a promising technology, than a detour or even a delay, will encounter much resistance (pp.171-172).

An important way in which actors create promises is by using formal methods for futures thinking. What do such methods do? Also here, the famous S-curve serves to illustrate the point. An S-curve enables an actor, for instance a firm, to position products and their promise (Van Lente 1993, p.85): '[b]y locating a product or process in one of the stages firms decide how much will be invested in it [and thus] S-curves provide incentives for action.' One could denunciate the manipulation of S-curves, like Galbraith and Merrill, and accuse actors of having hidden 'bad intentions.' It seems more fruitful, and more in line with the observations, to simply say that methods for strategic analysis are part of the process and that they perform within the process: they *act* (p.191). Viewed in this manner, they serve as *interessment* devices (Callon, 1986b) and hence reinforce actors' credibility cycles (Latour and Woolgar, 1986). Van Lente even gives a generic name to all formal futures (i.e. foresight, forecasting and strategic analysis) methods. Comparable to Vedin's 'intellectual technologies' (1990), he suggests to call them 'expectation technologies' since methods used in futures thinking help to create expectations. Expectation technologies are thus one type of technology amongst others that in practice structure innovation processes.

The literature discussed in this section aimed to provide the reader with a better understanding of the role of futures methods within innovation processes. Such methods are both embedded in and produced by such processes. At the same time, they help to shape and structure them. Since Giddens (1984) structuration theory sees (social!) actors as both embedded in and constitutive of social processes for sociologists the point may sound familiar. However, for futurologists this is a radically different interpretation than the one that initially framed the development of futures methods, namely, the expectation that they would allow us to *predict* the future.[13] In the interpretation given here it is not so relevant to know how well these methods do or do not predict future developments. The important thing to realise is that they are (or might be) performative and thus manage to bind actors together. Their results do not come true, but are eventually *made* true (a point also made in Chapter Two of this volume).

Even if this reasoning might sound plausible analytically, it does not make the task much easier. If futures methods and their results are constitutive of, and embedded in, innovation and decision processes, will the choice for a method or tool be arbitrary? One should not fall into a new positivistic trap, being to search for the method that would 'best' tie different actors together. Again there seems a parallel to be drawn with Forrest (1991 p.451), who writes that 'Managers of technological innovation and strategic planning should realise that it may be more effective if they develop models contingent upon their own situation'. In other words, management practices and models are local. Even if formal and general methodologies for strategic analysis exist, new practices should build upon existing ones. Thus, if one wants to 'develop models contingent upon one's situation' as Forrest indicates, and if we are interested in the issue of technological development, one needs an understanding of the underlying features of that development. These can be found in *innovation studies*.

Innovation Studies: Three Relevant Findings for Foresight and Strategic Analysis[14]

Developing tools for futures thinking in research and technology policy, requires a discussion of the findings from the study of technical innovation. Even if general theories differ enormously between, for example, sociology and economics, lines of consensus can be identified which have implications for thinking over future developments.

This section discusses three aggregate results from innovation studies and reviews their implications for futures thinking. The first relates to the observation that innovation is not a linear process from science to market. On the contrary, it consists of the gradual establishment of links between actors. Second, innovation processes consist of the formation of heterogeneous assemblies, being simultaneously social as well as technical. Lastly, innovations are path-dependent and can quickly become irreversible.

Innovation as an Interactive Process

Innovation has long been described as a stepwise process from science to the market.[15] Schumpeter (1934) was the first to analyse technical innovation processes more thoroughly and to conceptualise technical innovation in stages of invention, innovation and diffusion. During invention, new ideas (sketches, models) are generated. In innovation some of those inventions can be picked up by firms and brought onto the market. Finally, the diffusion phase consists of imitations of the original 'prototype' adopted by a larger and larger number of adopters. The reason for a technology not being adopted, even when profitable, would lie in the conservatism of the group of potential adopters. In sum, as in a relay race, actors in the innovation process supposed to deliver their products sequentially - first knowledge then a concept or patent, then a prototype, etc. As in the race, each withdraws having delivered their finished contribution. This process was supposed to be governed by a logic of either science-push or demand-pull.[16]

However, empirical studies show that innovations do not follow such a sequence. Rather, the objects developed and elaborated by actors describe a whirlwind pattern - as proposed in a review by Vinck (1991). Products, whether they are consumer products or scientific articles, are developed and tested, maybe normalised, but will also encounter resistance, be reshuffled, redefined, tested again, and so on. They can go from science to market, but they may also find their origin in a certain demand. In other words, scientists sometimes play a dominant role, but as Von Hippel (1988) showed, users can also be important in this process. Finally, invention can even take place in what normally is called the 'diffusion' stage. No one general factor, no one single social group can solely be held responsible for the development of an innovation.

Whereas economists have proposed new models to deal with the non-linear character of innovation,[17] the traditional linear conception of innovation still very often forms the basis for futures thinking. For instance, the Porter et al. (1991) handbook on foresight techniques and methods writes: '...the supply of scientific knowledge and technological principles [vs.] economic and societal demand...' and proposes a 7-stage, though still highly linear, innovation model (p.59) as '...general concept of how technologies develop...' (p.58). In contrast, innovation studies suggest that innovation is not a process whereby actors (laboratories, firms, users...) intervene sequentially, but one during which durable links are bit by bit created between these various players. As stated earlier, foresight exercises themselves, being embedded in innovation processes, are typically of such a collective nature. Symmetrically, futures thinking may want to account for the collective character of innovation.

The Social and the Technical are Simultaneously Shaped

The second insight derived from innovation studies concerns the relation between the technical and the social. The technical, and more generally the construction of so-called technical systems, has until recently been the privileged domain of the technologist, the scientist or the engineer. The social world and within that the study of interactions between human beings was the exclusive preserve of the sociologist, or maybe the behavioural psychologist. Micro-studies of innovation have however shown that the two are intimately linked.

Whether we look at bicycles or bakelite at the beginning of this century, a gasifier in a developing country or high-tech subway and aircraft projects,[18] in all cases we see that the social and the technical are shaped *together*. Techniques are shaped by collective social action, simultaneously shaping the politics, the users, and the infrastructure associated with them (Callon, 1987). And once established, techniques act in themselves (Latour, 1994). Hence, a technical object does not only reflect social relationships (Bijker, 1995) but also '...transcribes and displaces the contradictory interests of people and things...' (Latour 1992c). In sum, innovation - as a process as well as in terms of its outcomes - is a sociotechnical assembly in which multiple heterogeneous actors play a role. For the purpose of our enquiry it means that research priorities and technological options are always connected to certain configurations of actors - those through which they are constructed as well as those which are projected onto the future. Put another way, every technical choice implies or presupposes a social configuration, and *vice-versa*.

Irreversibilisation Occurs ... and Often Rather Early

The third implication for futures studies stems mainly from the work of economists of technical change and their observation that innovations are path-dependent. Dosi's, or Nelson and Winter's trajectories, or again Sahal's innovation avenues account for the existence of paradigms within technical innovation that are relatively fixed over long periods of time.[19] These paradigms are built up little by little, resulting in the end in a framework shared by everyone. As David (1986) showed for the QWERTY keyboard, such trajectories are not necessarily 'rational' or 'optimal' with regard to the technology under concern. They heavily depend on initial choices becoming (quasi-)irreversible.

David (1986) explains that in the case of QWERTY, the arrangement of the keyboard was the most optimal way to prevent the compacting of typebars òne behind another. For the manual act of typing itself more optimal solutions might have existed (by 1867 there were more than 50 patent attempts describing a commercial typewriter). Moreover, Remington was near to broke when it bought the manufacturing rights from the inventor. Nevertheless, schools for novice typists based themselves on the Remington machines. Likewise the early manuals. Consequently the supply-line of operators were trained in QWERTY, influencing buyers and suppliers. QWERTY's final standardisation in 1890 came from the expectations that the buyers of typewriting machines had with regard to the equipment users, who were now used to the lay-out. Despite the fact that in 1890 new machines had been developed with both a better key arrangement and a better way of preventing the typebars from compacting it was already too late: now, some 100 years later, the greater part of the world still finds itself typing on a QWERTY keyboard. For the other parts a symmetrical story can certainly be told (France for instance uses 'AZERTY'-type keyboards). Note that observations of the sociologists cited in the preceding paragraph and the economists cited here converge: the QWERTY-story shows very well that the simultaneous stabilisation of both social and technical elements accounts for irreversiblilisation and path dependency of innovations.

Other authors argue that stabilisation occurs early, fostering the development of a certain trajectories and excluding others. Small but critical events appear important in any technical development, whether this is within research programmes (Rip, 1995), in the case of products competing for a market share (Arthur, 1989) or for technological systems (Cowan, 1990). Hence the third consequence for foresight studies is the need to identify those areas where early closure of sociotechnical choices is at stake. Methods such as regression analyses, logistic - 'S' - curves, envelope curves, or other types of extrapolation can indeed be used to describe trajectories but they eventually will have predictive power only *after* irreversibilities have been created.[20] Before, they are focusing devices, which serve to create expectations that may bit by bit gain credibility.

Consequences for Foresight and Strategic Evaluation

The implications for strategic analysis and foresight of the three closely related findings of innovation studies sketched above can be summarised as follows.

First, the observation that innovation is an interactive process means that in principle nothing should be viewed as leading inevitably to an innovation. Rather, innovation is the result of the gradual establishment, through trial and error, of links between scientists, engineers, users, firms, and so forth. Consequently, markets, if ever they come into being, are a result of this sociotechnical enterprise rather than a given. Although one may find that Garcia (1986), Granovetter (1985) or, long before that, Polanyi (1944) may too much 'sociologise' the coming into being of markets (the latter as a social constructivist *avant la lettre*), their common argument is clear: markets do not exist *a priori* but are the tangible result of a process that cannot exist without the social relationships they are embedded in.

The implication for foresight is as follows. Scientific findings cannot be expected to automatically, or even necessarily, find their way to the market. The idea that successful technology responds to market conditions (Porter et al, 1991) has no meaning for the futures thinker, since most of the time such markets (as well as the technology, the science and all other elements which have to be in place) will have to be constructed. Also users will have to be interested in order to do so. Thus, future methods should not presuppose the existence of markets - nor of social needs - but should instead take their construction as its first point of departure. This could be done by explicitly accounting for the fact that innovation is an interactive process involving actors. As outlined above, it is also important to remember that foresight is itself a product of such processes.[21] These observations notwithstanding, it is important to see that actors in often need to 'linearise' future technological developments in order to design their strategies (Schaeffer, 1998).

The second finding presented above states that innovation is social as well as technical. This necessarily means that every technical choice is simultaneously a social one – whether it is implicitly or explicitly made. A technical choice distributes roles to actors, and makes hypotheses about the environment in which the technical item is to be inserted. Within the framework of technology foresight or strategic analysis, this means that technological options can no longer be viewed as technical matters only but should always be viewed as *sociotechnical* options. 'Technology' Foresight is in all respects sociotechnical foresight.

The third lesson was that lock in and path-dependency occur within innovation processes. Such points of irreversibility are highly dependent upon historical events and initial choices. This finding has severe consequences for technology management, since it means that certain actors persistently try to privilege one trajectory, whereas hopes of development in this direction are in fact already lost. Without anyone even noticing it yet, the privileged route may have been overruled by another trajectory. And once this latter trajectory has matured it will be difficult to replace it by another one, at least in the short term. The importance of this insight for futures thinking is obvious. As mentioned above, one can have severe doubts about the use of extrapolative methods in situations in which nothing has yet stabilised. Such methods might at best be seen as part of a strategy adopted by actors seeking to push forward the trajectory they want to privilege. On the other hand, new foresight methodologies, starting from this observation, should focus on detailed investigation of when and where irreversibilisation might occur or has already taken place.

Futures Methods and Sociotechnical Networks

Viewing innovation processes in terms of the construction of sociotechnical networks allows us to integrate the three findings of innovation studies outlined above. First, the notion of network refers both to the actors that make it up and simultaneously to the intermediaries that circulate between them - it thus accounts for the interactive character of innovation. Second, the notion of network accounts for innovation as being sociotechnical not only because it describes the association of heterogeneous actors - belonging to the scientific or technical sphere, industrial production, commercial distribution, consumption, administration or other - but also because co-ordination mechanisms can be those traditionally described by economics,[22] as well as those linked to science and technics (texts, embodied knowledge and objects: Callon, 1992a, 1992b, 1995). Third, it is the gradual coming into being of such coordination which accounts for the stabilisation and irreversibilisation of sociotechnical networks (Callon, 1991). Thus, a theoretical distinction can be made between fully stabilised networks, and those which are completely unstable. In this perspective, a perfect market represents a stabilised network (but there might be other such theoretically stabilised forms). Science, inversely, is more in the business of having new networks emerge - by producing new ones all the time: it creates diversity (Callon, 1994).

Processes of research and technical innovation can therefore be viewed as processes in which actors propose, test and try to realise *new* trajectories and try to stabilise certain *sociotechnical networks*. Consequently, they make the hypothesis that a future, corresponding to 'their' trajectory, can be realised on the futures which might/should be created - and exclude others from their field of action. Outcomes of futures methods are used in such processes to promote trajectories, and reduce the 'value' of trajectories proposed by others. Viewed in this manner, technology foresight may become a *forum* in which actors confront the future scenarios they elaborate through the networks they try to construct. It is a forum, since it is an open space in which actors are confronted with each other and debate choices which involve the collectivity. It is a *hybrid* forum (Callon and Rip, 1992), since the identity of the actors is not fixed beforehand, nor are criteria for choice between scenarios.

From this point-of-view, potential future trajectories, successful or not, *can be viewed as being inscribed by actors in their attempts to stabilise new sociotechnical networks or modify existing ones*. A laboratory focusing on gene mapping, a public research programme financing soil cleaning techniques, the writer of computerised simulation models for rock mechanics: all anticipate a specific network in which their results are to be taken up. In the first case, a world is shaped where a market for genetic techniques exist. In the second case, general agreement exists to define and locate contaminated soils that have to be cleaned up. In the third case the idea might be to exploit oil wells more efficiently to keep our cars on the road as long as possible. These worlds, or at least some parts of them, are inscribed in the routines used, the articles put into circulation, the instruments developed, the labs put together or the industrial partners involved. Any research project promotes a specific sociotechnical network and therefore anticipates a world that may or may not be realised. It *endogenously* anticipates tomorrow's world, which all those who support the project share (explicitly or implicitly).

In such an approach, the tasks of a futures analyst is to identify which trajectories are anticipated to the different technological project proposed by actors, and how these can be compared. The following and last section will give some tools I have developed in the past to do so.

Two Complementary Foresight Tools Derived From Innovation Studies

Over the last few years colleagues and I have been involved in working with managers of technological programmes and with researchers to make the principles sketched above useful for concrete research programme management. Several tools were developed to help in comparing and, hopefully, designing, new research and technological programmes.[23] I do not pretend that these will by themselves be able to guide future directions. Exactly as with traditional futures methods, humans are needed to interpret them and take up their results and make collective choices. They will have to be embedded in the process of strategic decision making and their results should constitute points for debate. However, feeding some basic notions from innovation studies into strategic decision making in technology, has proven to be a great help. The contribution of the tools presented below lies especially in their ability to compare the societal content of research programmes, and to make managers and researchers aware of the fact that the choices they make are more than simply technical answers to technical problems: they also define a new world which has to be simultaneously built up.

Making Future Scripts Explicit

National research funding agencies and research programmes are a typical locus of translation of (national) goals into implicit and explicit sociotechnical choices. As from 1992, the French energy and environment agency ADEME was prepared to be our 'guinea pig' and helped us in developing new tools. This agency is active in the field of non-nuclear energy, air pollution and waste management, and, amongst other things, funds research in all these areas.

Sociotechnical analysis describes objects and the confrontation with their environment 'on site' or retrospectively. In analysing the research programme of a public agency however, we are confronted with objects that, albeit in construction, exist in most cases as *text* only: on paper, or incorporated into interviewees. Sometimes they happen to be partially objectified in prototypes, patents or demonstration programmes. Their 'description' (Akrich, 1992) will most of the time have a fictive character and therefore depends on thought experiment.

In order to conduct such thought experiment, we made use of Akrich's concept of 'script.' Every existing technical object contains a script (in words), which attributes roles to social as well as technical actors. It defines what the actors are, do and want. For example, a cash dispenser is typically in command of its user: put your card in the machine, type your secret code. It defines a connection to your bank account and thus to your bank, it checks whether the account contains enough money and will then - hopefully - decide to agree with the transaction you expect from it. It defines a complete environment which is need in order to function, and at the same time it acts itself upon this environment.

In the case of objects that are promoted through research - and therefore do not exist as yet - such an analysis is not so easy. At least three thought experiments can serve to make the future object's 'expectations' explicit. They all come down to finding ways of establishing the list of (both social and technical) actors the object presupposes and the roles these may be forced to play. The first consists of roughly describing the world as it will be if the service that corresponds to an end-product proposed through a piece of research would be entirely delivered by that product only - we will call this the *'100% thought experiment'*. This thought experiment allows us to quickly grasp the general characteristics of the object in question, and especially the 'barriers' existing in today's world to realise the object of concern. If for instance the object pursued would be a battery for an electric vehicle, the '100% thought experiment' would lead to the statement that all gas stations should deliver electricity - which today is typically not the case.

The second thought experiment goes one step further. It consists of *black-boxing* the object as it is expected (by an engineer for instance) to function in the future and analyse it as an 'input-output device.' This makes it possible to identify the actors connected to those in- and outputs. The third thought experiment concerns the analysis of the relationships between actors incorporated into the object: what does the object pre-scribe, oblige, allow, forbid, stimulate, etc. Although this requires a rather detailed analysis of, for instance, concrete project descriptions, it is specifically well fitted for describing what is expected from future *users*. We can again the electric vehicle as an example. If its battery is designed as a single module, it may be rapidly exchanged for a new, fully charged, one. If however it is divided over different places in a car (which for instance was the case with the first electric Peugeot's 106) it necessarily will have to be recharged at home or at an 'electricity station'.

The three, complementary, thought experiments provide a preliminary description of the future world the object defines. It makes comparison with other objects in construction, and thus different anticipated futures, possible.

The formalisation of the implicit and explicit hypotheses that are built into the objects promoted by the agency whose actions we analysed was not easy to achieve since they differ from case to case. We will shortly discuss an example to illustrate the 'black-box thought experiment.' It concerns different routes to deal with plastic wastes, envisaged, in 1993, by the agency. Sometimes, a great role is given to infra-structural elements (for example 'a technical system for collecting plastic waste'); in other cases future users are asked to change their behaviour ('please separate plastics manually from your other types of household waste'). Three different routes for dealing with plastic wastes were considered by ADEME in 1993: mechanical separation, chemical recycling (based on cracking), and incineration, the traditional way of waste disposal. As a thought experiment, we will consider the three as being black-boxed and stabilised (where by the time of our investigations only incineration more or less was), which allows for an analysis of what futures they define according to our interviewees. One can then establish the following (simplified) table.

Table 9.1 Three different futures competing to realise the same promise

Expected technical object from research project	Separation technology	Chemical recycling technology	Incineration technology
Agency's overall goal (cf. Van Lente's 'promise'), same in all three cases	Recycling of plastic household waste		
Expected output of future object	Separated plastics of high purity	Naphta, synthesis gas	Heat, electricity
'Downstream' actors (i.e. associated with the output)	Plastics recycling industry (heterogenous, mainly SMEs)	Chemical industry (oligopolistic, few firms)	Consumers of energy (diverse); landfill
Expected input to future object	Relatively clean mixed plastics from household wastes	Mixed plastics from household wastes	Mixed wastes
'Upstream' actors associated with the input	Separate plastic waste collection & supply => households or machines that separate plastic waste from other waste	Separate plastic waste collection & supply => households or machines that separate plastic waste from other waste	Mixed waste collection & supply => households that act as they were used to (all waste in the same bin)

We have analysed all the fictive scripts of the research activities of the national agency. This made it possible to describe the future scenarios the agency *de facto* was constructing, as well as the ones it did not pay attention to. The total set of scenarios thus established can be called the 'prospective profile' of the agency (De Laat, 1998). Of course, apart from national agencies, the method can apply to any other type of actor by assembling the fictive scripts projected onto the future by the R&D projects promoted by that actor.

Confronting De Facto Scenarios of National Agencies Involved with Research

Studying the future worlds promoted by the French agency, raised the question to what extent they related to worlds that were pursued by *other* actors pursuing similar goals, and whether it was possible to compare them to these worlds. On selected fields, we investigated the research promoted by NOVEM, the Dutch energy agency and ADEME's counterpart in the Netherlands. It quickly became clear that to respond to similar objectives, different and sometimes entirely opposite routes were chosen by the Dutch agency. A full account of this comparison is given in De Laat and Larédo (1998). Here only major results will be recalled.

As has been shown in the table in the example concerning plastic waste, a shared overall goal can lead to different technical solutions, and therefore, to different anticipated *sociotechnical* worlds. The comparison between the two agencies showed that such future scenarios relate to each other in different configurations. Names from traditional economics were borrowed to 'label' these configurations: 'monopoly,' 'oligopoly,' 'monopsony,' and 'pure competition.' These labels should not be taken in their economic meaning, but as metaphors for the configuration in which different 'future worlds' take it up against each other.

Hence, in a *monopoly situation*, several, often still weakly developed, future worlds face a dominant existing world. Though today this seems to be slightly changing, this configuration was encountered for nuclear electricity production in France, for which ADEME, involved with *non*-nuclear energy only, proposed alternative *de facto* scenarios. However, because of the broad diffusion of nuclear energy in France (today still nearly 80% of electricity is generated by nuclear power) the agency sought to build new networks in very specific places only. For instance, research on wind power was promoted, but demonstration projects took place only in overseas departments not provided with nuclear electricity. Also development of solar cells was funded, but applications were mainly found in remote sites in the mountains, where it would be too expensive to bring nuclear electricity to. Small hydropower constituted a similar case. In other words, the agency followed a niche strategy in order to occupy the few spaces left open by nuclear, which, referring to Hughes, in France is typically *the* network of power (Hughes, 1983).

A very different configuration was given by controversies playing around biomass in the two countries analysed. To simplify, both in the

Netherlands and in France actors tried to establish translations of the same problem, namely the greenhouse effect involving carbon dioxide emission, and the 'Set aside policy' of the European Union concerning agriculture. In both countries, these problems were globally translated into the development of so-called energy crops, since these fix carbon dioxide and occupy fallow land. They are therefore interesting from the point of view of both energy and agricultural policy.

Although the general problem definition from which the two agencies started seemed identical, the technical solutions proposed were radically different. After a series of economic studies, the Dutch agency proposed to the Government to gasify trees for electricity production, whereas the French, after a long political debate, had opted for the biofuel route. Fictive script analysis shows that the consequences of these two routes are not the same. The future agricultural world associated with the biofuel route concerns (traditional) cultures like wheat, beets, sweet sorghum or rapeseed, and the products it delivers are expected to substitute for fossil fuels, for (traditional) combustion engines, in transport. In fact this world looks a lot like the one we know today. The gasification route however involves growing trees, which means that farmers, now confronted to cycles of 6 to 10 years or even more, are expected to change not only their harvesting practices but should also radically restructure their accounting habits! The economic feasibility study thus overlooked the social consequence of what seemed technically a perfectly justified choice. And indeed, the criteria chosen for the study were not accepted - neither by the Ministry of Agriculture, nor by the Farmer's Association.

Whereas the Dutch energy agency basically thought that with their report the case would be closed, in reality it started a typical case of technological controversy leading to the definition of a research programme investigating, amongst other things, the social consequences of different biomass options. We have termed this configuration of competing networks between France and The Netherlands an *'oligopoly situation'*: two or three complete and mutually exclusive scenarios compete to provide a *full* solution to the same problem but there is no accepted method or procedure for relative comparison since networks are not stabilised.

In contrast, the third configuration is based on an end-state shared between actors, but on partly different paths - which actors try to build *within* the boundary conditions given by the end-state, itself the result of political debates. This was the case with ADEME's waste scenarios. The shared end-state was that, by 2002, all waste should as much as possible be recycled, and only 'ultimate waste' should be disposed of in (protected) landfills. The critical issues therefore were (1) gaining consensus over the definition of ultimate waste and (2) the organisation of the separation, collection and recycling of waste which would fall outside this definition. The latter point, as in any innovation, required a lot of trial and error processes in French municipalities and firms, through which learning took place. The objective was to progressively align all participants by constructing equipment specially adapted to household waste, and the establishment of new services. Though locally systems were very different,[23] all efforts pointed at a similar end-state where waste was transformed into a 'mine' of new materials entering nationally or internationally established markets (paper, aluminium, iron, glass, etc). Referring to the economic situation in which one client has different suppliers (e.g. as in the case of a national railway), the metaphor of the *monopsony* was chosen for this configuration: 'one' shared future world is constructed through different, local, routes.

In the fourth situation - pure competition - future worlds proliferate. All actors suggest different worlds of tomorrow, and different frames of reference correspond to each individual world. Positions are often unique and few elements are shared. Whereas in the first three cases networks have to be stabilised to a certain extent, and policy options can often be expressed, in the fourth case, inspired by a survey on long term environmental problems, networks are not configured at all. Futures study can only provide for an inventory here. It is the situation where 'foresight' is most needed, but, since it is unstable, it is the situation where foresight is least helpful to predict.

This typology of possible configurations has been used as a tool to characterise and compare the configuration of future worlds proposed by different actors. In practical innovation management they assist in the analysis of the type of intervention one can adopt given a certain configuration or indicate the type of controversy one is involved in. The different configurations are summarised in the table below.

Table 9.2 Four situations of scenario-confrontation (after De Laat and Larédo, 1998)

Type	Description	Types of controversy	Actions encountered
Monopoly	Every actor promoting new options is confronted by a stabilised, dominant and central techno-economic network	A battle 'of minorities against the majority'	Niche strategies Price competition
Oligopoly	A limited set of radically different scenarios are proposed by different actors in response to the same problematic; trajectories and end-states differ; no criteria have yet been constructed which would allow for a choice	Controversy over the criteria for choice and over the organisation of society ('social content') associated with technological options proposed	Political process which allows the construction of criteria for choice between different networks
Monopsony	All networks proposed by actors postulate the same end-state, but proposed trajectories differ	Eventually, 'traditional' scientific and technical controversy	Demonstration programmes along existing criteria
Pure competition	Represents a situation in which all actors propose different scenarios as well as different criteria to choose between them.	In principle, possible about everything since nothing is stabilised	Inventory of the futures proposed by actors

Conclusion

This paper argued that traditional futures methods used in decision making around technological research often tell us only half of the story. They tell us which macro-evolutions or scientific breakthroughs we can expect and may also warn us about negative impacts of new technologies. Whereas futures methods help us to describe the *promise* of a technology (all the problems it will help us to solve...), they hardly tell us anything about what the world should look like in which these technologies are expected to live. I claim that 25 years of innovation studies allows us to view technological options proposed by actors as attempts to stabilise different future trajectories. One of the tasks of the *futurologist* is to help in making explicit these trajectories, and compare the future worlds they may lead to. I have presented some simple tools to do so.

In order to accept this job, 'futures methodologists' should agree with me on two major points. The first concerns the status of futures studies within innovation processes, the second concerns the way in which to view innovation. As concerns the status of future studies (including forecasting, foresight, strategic analysis, etc.) this paper makes a long analytical detour in order to show that future studies and their results are both embedded in, and constituent of, decision making processes. Consequently, instead of *predicting* the future - which was typically the way in which futures methods were viewed when they were first developed - they help, as one ingredient amongst many others, to build the future or push the present into one direction instead of another, a point also made by Deuten and Rip elsewhere in this book.

The futurologist will need to adopt a view on innovation as being an *interactive* process in which many heterogeneous actors intervene. This process is not going linearly from science to the market, or from basic research to applications, but instead is characterised by multiple to-ing and fro-ing between such 'stages.' In this process, irreversibility may occur very early, determining the range of possible choices left open. Finally, the prospective nature of technical innovation lies in the idea that every technical choice is simultaneously a choice for a certain 'social' organisation. This has led to the definition of the concept of 'fictive' or 'future' script, allowing us to describe the *future* actors and their relationships defined through *current* technological research. On the basis of this concept, several tools have been developed to make implicit scripts of research projects explicit, to collate them into *de facto* scenarios, and to compare these amongst each other. The results of innovation studies give useful insights to design new foresight tools.

So far, these tools have been used mainly on a fairly operational level. They have been used by research funding agencies, as a complement to traditional methods in decision making regarding their programmes. But they also appear to be very well suited for educational purposes, to teach future innovation managers about the 'social side' of their work. However, the ultimate value of being able to compare the future worlds defined through research lies elsewhere. If the comparison of different research *programmes* yields a future scenario *confrontation*, this may indicate that options proposed are not only technically incompatible, but define radically different worlds for which neither science (including economics) nor futures studies is able to propose selection criteria. Although the boundaries are not always that clear, technical controversy may best be fought out between scientists or engineers. However, if research programmes are proposed which involve entirely new social, economic, and other, relationships, choice is no longer a technical issue. In such a case, the choice between different pieces of research becomes a choice between ways we want to organise future society. For this reason, the analysis of fictive scripts and the comparison of *de facto* scenarios is, first of all, political.

Notes

1. Hoos, cited in Linstone, 1989, p.2.
2. Latour and Woolgar, 1986 (1979). For an overview of the critiques (and a critique itself) of social constructionism see Michael, 1996, especially Ch.3.
3. Or a complete loss of foundation - see, for instance, Collins and Yearley, 1992, pp.322-324. For some counterarguments see Ashmore, 1989; Callon and Latour, 1992; Woolgar, 1992, p.331; or Ashmore, Edwards, and Potter, 1994.
4. For this argument he draws upon Geertz's interpretative anthropology (Geertz, 1983, 1993 [1973]) and to a lesser extent on other humanities like literary criticism, semiotics and post-critical sociology. The importance and relevance of 'cultural' differences seem to become more and more recognised in the technology management literature.
5. For overviews see Cetron, 1969; Jantsch, 1967; Linstone, 1989; Linstone, 1994; Linstone and Turoff, 1975a; Porter, et al., 1991; Twiss, 1994.
6. De Laat, 1996.
7. In the energy field for instance; see Blok, et al., 1995.
8. As in Meyer, et al., 1993 or Den Hond and Groenewegen, 1996 who situate foresight on a 'sectoral' level.
9. Drejer, 1996 is original in his dividing different approaches to management of technology (MOT) in four schools of thought. This allows him to distinguish the different methods used by different scholars:

School 1: R&D Management	technology forecasting and budgeting
School 2: innovation management	delphi forecasting, technology forecasting, project management of the innovation process
School 3: technology planning	scenario forecasting, technology analysis and planning
School 4: strategic MOT	strategic MOT, organisation-technology approach to MOT and integrated MOT

Although this table makes clear that there are differences in uses of methodology that are closely linked to the view of technology management one adopts, the boundaries are still blurred and non-specific. The author does not account for the relations between different schools – and thus between methods – which may not exist in academia, but do exist in practice, i.e. in organisations managing their technology. In the case that Delphi exercises are used as input to scenario development for instance boundaries between methods, and consequently, the author's schools of thought, and thus clear distinctions between strategic analysis, foresight and forecasting, vanish.

10. Neither should firms, as indicated by Coates: 'Any forecast embraced by the organization has to be reworked, re-evaluated and interpreted for its implications for the organization. One would be ill-advised to take any forecast lock, stock and barrel and do one's planning based on it. It must be evaluated and interpreted for its implications for the user' (Coates, 1995 p.12).

11. Cf. p.119, 'Leach has basically talked to a lot of knowledgeable people, and he has selected opinion from those people which lead to minimum energy growth.' This person, named 'Leach' was thus seen by one of his colleagues as being far too optimistic and De Man uses this as an argument in favour of the social construction of forecasting methods (although not labelling it as such), since by taking other 'knowledgeable people' to do its research Leach would have come to other conclusions - cf. the editor of the journal Energy Policy accusing Leach's book of being 'heretic[al]' (p.122).

12. Cf. p.88: 'The public energy policy discussion had ended in a political vacuum. It seems unlikely that energy scenarios will again come to play an important role in the near future.' Although the author clearly expresses the fact that at the moment he conducted his analyses the energy scenarios he discusses were not adopted, his own medium term forecast appears to be wrong: energy scenarios have since undergone a very important come back, for instance through the research funded by the former European programme JOULE.

13. See Linstone and Turoff, 1975a.

14. An earlier version of this discussion can be found in De Laat and Larédo, 1998.

15. See Freeman, 1982 or Forrest, 1991 for overviews.

16. Coombs et al., 1987 give an extensive overview of the technology push-demand pull debate.

17. Cf. the chain-link model of Kline and Rosenberg, 1986.

18. See, respectively, Akrich, 1993; Bijker, 1987; Latour, 1993a; Law and Callon, 1992.

19. Dosi, 1982; Nelson and Winter, 1982; Nelson and Winter, 1977; Sahal, 1985. Whereas they focused on typical products/technologies, such regimes are said to exist on the level of generic technologies Perez, 1983, countries Lundvall, 1988 or social groups Bijker, 1987 as well.

20. One could argue that irreversibilisation also determines our notion of time (again another important issue for foreşight!), flowing in one direction. It is this which makes, for instance, a recent novel by Martin Amis seem so unreal (though highly interesting). In this novel, the storyteller starts at the end of the main character's life and then walks back, i.e., towards the main character's youth. This is not done in the more common form of a flashback, instead the narrator systematically experiences the main character's life in reverse order (which in Jansen's terms would be an extreme form of *backcasting*; see Jansen, 1991). Consequently, he sees things undone all the time, and become non-existent, without leaving *any* trace (relationships for instance). The fact of the main character being a doctor comes then as an extra surprise since, if time is put upside down, the doctor makes people ill instead of healing them.
21. This argument has been worked out in detail for the method of cost-benefit analysis by Porter, 1995, ch.7.
22. I.e. markets, hierarchies (Williamson, 1996), and trust (Karpik, 1996).
23. The European project SOCROBUST (programme TSER) for example will use some of the tools discussed here with other instruments in view of managing European radical innovations. The Dutch energy agency is currently using the idea of future script in order to design new technology programmes.

References

Akrich, M. (1992) The De-scription of Technical Objects, in Bijker, W.E. and Law, J. (eds), *Shaping Technology/Building Society: Studies in Sociotechnical Change* 205-224, , MIT Press, Cambridge, Mass.

Akrich, M. (1993) Essay of Technosociology: A Gasogene in Costa Rica, in Lemonnier, P. (eds), *Technological choices. Transformation in material cultures since the Neolithic*, 289-337, Routledge, London.

Amara, R. (1989) A Note on What We Have Learned About the Methods of Futures Planning, *Technological Forecasting and Social Change*, 36, 1-2, 43-47.

Arthur, W.B. (1989) Competing Technologies, Increasing Returns, and Lock-in by Historical Events, *The Economic Journal*, 99, 116-131.

Ashmore, M. (1989) *The Reflexive Thesis. Wrighting Sociology of Scientific Knowledge*, The University of Chicago Press, London.

Ashmore, M., Edwards, D., and Potter, J. (1994) The Bottom Line: The Rhetoric of Reality Demonstrations, *Configurations*, 1 (1) 1-14.

Bijker, W. (1987) The Social Construction of Bakelite; Towards a Theory of Invention, in Bijker, W.E., Hughes, T.P. and Pinch, T.J. (eds) *The Social Construction of Technological Systems*, 159-187 MIT Press, Cambridge, Mass.

Bijker, W.E. (1995) *Of Bicycles, Bakelites and Bulbs: Towards a Theory of Sociotechnical Change* MIT Press, Cambridge, Mass.

Blok, K., Turkenburg, W.C., Eichhammer, W., Farinelli, U. and Johansson, T.B. (1995) *Overview of Energy RD&D Options for a Sustainable Future* (joule report No. JOU2-CT 93-0280; EUR 16829 EN). European Commission, Directorate-General XII, Science, Research and Development.

Bowonder, B. and Miyake, T. (1993) Technology Forecasting in Japan, *Futures*, 25 (7) 757-

776.

Bright, J. R. (1986) Improving the Industrial Anticipation of Current Scientific Activity, *Technological Forecasting and Social Change*, 29 (1) 1-12.

Callon, M. (1986a) The Sociology of an Actor-Network: The Case of the Electric Vehicle, in Callon, M., Law, J. and Rip, A. (eds) *Mapping the Dynamics of Science and Technology*, 19-34, The Macmillan Press, Houndsmill.

Callon, M. (1986b) Some Elements of a Sociology of Translation: Domestication of the Scallops and the Fishermen of St Brieuc Bay, in Law, J. (ed) *Power, Action and Belief. A New Sociology of Knowledge*, 196-233, Routledge and Kegan Paul, London.

Callon, M. (1987) Society in the Making, in Bijker, W., Hughes, T. and Pinch, T. (eds) *The Social Construction of Technological Systems* MIT Press, Cambridge, MA.

Callon, M. (1991) Techno-economic Networks and Irreversibility, in Law, J. (ed) *A Sociology of Monsters*, 132-161, Routledge and Kegan Paul, London.

Callon, M. (1992a) The Dynamics of Techno-economic Networks, in Coombs, R., Saviotti, P. and Walsh, V. (eds) *Technological Change and Company Strategy*, 73-102 Academic Press, London.

Callon, M. (1992b) Variety and Irreversibility in Networks of Technique Conception and Adoption in Foray, D. and Freeman, C. (eds) *Technology and the Wealth of Nations*, 275-324, Frances Printer, London.

Callon, M. (1994) Is Science a Public Good? Fifth Mullins Lecture, Virginia Polytechnic Institute, 23 March 1993, *Science, Technology and Human Values*, 19(4) 395-424.

Callon, M. (1995) Technological Conception and Adoption Network: Lessons for the CTA Practitioner, in Rip, A., Misa, T.J. and Schot, J. (eds) *Managing Technology in Society: The Approach of Constructive Technology Assessment*, 307-330, Pinter, London.

Callon, M., and Latour, B. (1992) Don't Throw Out the Baby with the Bath School! A Reply to Collins and Yearley, in Pickering, A. (ed) *Science as Practice and Culture*, 243-368, The University of Chicago Press, Chicago.

Callon, M., and Rip, A. (1992) Humains, Non-humains: Morale D'une Coexistence, in Theys, J. and Kalaora, B. (eds) *La Terre Outragée. Les Experts Sont Formels!* 140-156, Editions Autrement, Paris.

Cameron, H., Loveridge, D., Cabrera, J., Castanier, L., Presmanes, B.,Vasquez, L., and Meulen, B.v.d. (1996) *Technology Foresight: Perspectives for European and International Co-operation Final Report* No. PREST / EU.

Cetron, M.J. (1969) *Technological Forecasting, A Practical Approach*, Gordon and Breach Science Publishers Inc, New York.

Coates, J.F. (1989) Forecasting and Planning Today Plus or Minus Twenty Years, *Technological Forecasting and Social Change*, 36 (1-2) 15-20.

Coates, J.F. (1995) How To Recognise a Sound Technology Forecast, *Research Technology Management*, September-October, 11-12.

Coates, J.F., Mahaffie, J.B., and Hines, A. (1994) Technological Forecasting: 1970-1993, *Technological Forecasting and Social Change*, 47(1) 23-33.

Collins, H.M., and Yearley, S. (1992) Epistemological Chicken, in A. Pickering (eds) *Science as Practice and Culture*, 301-326, The University of Chicago Press, Chicago.

Coombs, R., Saviotti, P., and Walsh, V. (1987) *Economics and Technological Change*, Macmillan, London/Basingstoke.

Cowan, R. (1990) Nuclear Power Reactors: A Study of Technological Lock-In, *Journal of Economic History* (3).

Crookall, D. (1995) A Guide to the Literature on Simulation/Gaming, in Crookall, D. and Arai, K. (eds) *Simulation and Gaming across Disciplines and Cultures, ISAGA at a Watershed,* 151-177, Thousand Oaks, SAGE, London.

David, P.A. (1986) Understanding the Economics of QWERTY: The Necessity of History, in Parker, W.N. (ed) *Economic History and the Modern Economist,* Blackwell, Oxford.

De Laat, B. (1996) Suivre les Acteurs D'une Main, les organiser de L'autre. L'adoption de la Notion D'hybride par les Acteurs Eux-mêmes. In Rabeharisoa, V. and Méadel, C. (eds) *Coordonner, Attribuer, Représenter - Journées CSI, (9-10 mai)* 129-138, Ecole des Mines de Paris, CSI.

De Laat, B. and Larédo, P. (1998) Foresight for Research and Technology Policies: from Innovation Studies to Scenario Confrontation, in Coombs, R., Green, K., Richards, A. and Walsh, V. (eds) *Technological Change and Organization,* 150-179, Edward Elgar, London.

De Man, R. (1987) *Energy Forecasting and the Organization of the Policy Process,* University of Amsterdam, Delft, Eburon.

Den Hond, F. and Groenewegen, P. (1996) Environmental Technology Foresight: New Horizons for Technology Management, *Technology Analysis and Strategic Management,* 8 (1) 33-46.

Dosi, G. (1982) Technological Paradigms and Technological Trajectories, *Research Policy,* 11 (3) 147-162.

Drejer, A. (1996) Frameworks for the Management of Technology: Towards a Contingent Approach, *Technology Analysis and Strategic Management,* 8 (1) 9-20.

Forrest, J.E. (1991) Models of the Process of Technological Innovation, *Technology Analysis and Strategic Management,* 45, 439-453.

Freeman, C. (1982) *The Economics of Industrial Innovation,* Francis Pinter, London.

Galbraith, C.S., and Merrill, G.B. (1996) The Politics of Forecasting: Managing the Truth. *California Management Review,* 38, 2 (Winter) 29-43.

Garcia, M.-F. (1986) La Construction Sociale d'un Marché Parfait: Le Marché au Cadran de Fontaines-en-Sologne, *Actes de la Recherche en Sciences Sociales,* 1-13.

Geertz, C. (1973) [1993], *The Interpretation of Cultures,* Fontana Press, London (1st edition Basic Books, New York).

Geertz, C. (1983) *Local Knowledge: Further Essays in Interpretative Anthropology,* Basic Books, New York.

Georghiou, L. (1990) Evaluation of Research and Technology - Some Broader Considerations, in Krupp, H. (ed) *Technikpolitik angeschichts der Umweltkatastrophe,* 225-232, Physica Verlag, Heidelberg.

Georghiou, L. (1995) Participation in the United Kingdom Technology Foresight Programme - draft, in *Transformation of Organisation, 23-24 March,* 24+ill, Tilburg University.

Georghiou, L. (1996) The UK Technology Foresight Programme, *Futures,* 28 (1).

Giddens, A. (1984) *The Constitution of Society: An Outline of the Theory of Structuration,* Polity Press, Cambridge.

Godet, M. (1985) *Prospective et Planification Stratégique,* Economica, Paris.

Godet, M., Bourse, F., Chapuy, P. and Menant, I. (1990) *Problèmes et Méthodes de Prospective, Boîte à Outils.* Futuribles / ADITECH, Paris.

Gordon, T.J. (1989) Futures Research: Did It Meet Its Promise? Can It Meet Its Promise? *Technological Forecasting and Social Change,* 36 (1-2) 21-26.

Granovetter, M. (1985) Economic Action and Social Structure: The Problem of Embeddedness, *American Journal of Sociology*, 91, 3 (November) 481-510.

Hauptman, O. and Pope, S.L. (1992) The Process of Applied Technology Forecasting. A Study of Executive Analysis, Anticipation and Planning, *Technological Forecasting and Social Change*, 42, 193-210.

Hughes, T.P. (1983) *Networks of Power: Electrification in Western Society, 1880-1939*, Johns Hopkins University Press, Baltimore.

Irvin, D.J. (1993) Technology Forecasters - Soothsayers or Scientists, *IEEE Technology and Society Magazine* (Spring) 10-17.

Irvine, J. and Martin, B. (1984) *Forecasting in Science: Picking the Winners*, Francis Pinter, London.

Jansen, L. (1991) *Inaugurele Rede TU Delft*, University of Delft.

Jantsch, E. (1967) *La Prévision Technologique (Technological Forecasting in perspective)*, OECD, Paris.

Karpik, L. (1996) *Confiance et Crédibilité de L'échange Economique. Version provisoire* No. CSI Ecole des Mines (miméo).

Kline, S.J. and Rosenberg, N. (1986) An overview of Innovation, in N.A.O. Engineering (ed) *The Positive Sum Strategy: Harnessing Technology for Economic Growth* The National Academy Press, Washington, D.C.

Latour, B. (1993a) Ethnography of a 'High-Tech' Case: About the Aramis Case, in Lemonnier, P. (ed) *Technological Choices -Transformations in Material Culture since the Neolithic*, 372-398, Routledge and Kegan Paul, London.

Latour, B. (1993b) *La Clef de Berlin et Autres Leçons d'un Amateur de Sciences*, La Découverte, Paris.

Latour, B. (1994) On Technical Mediation - Philosophy, Sociology, Genealogy, *Common Knowledge*, 3, 29-65.

Latour, B. and Woolgar, S. (1986) [1979] *Laboratory Life: The Construction of Scientific Facts*, Princeton University Press, Princeton.

Law, J. and Callon, M. (1992) The Life and Death of an Aircraft: A Network Analysis of Technical Change, in Bijker, W.E. and Law, J. (eds) *Shaping Technology/Building Society. Studies in Sociotechnical Change*, 21-52, MIT Press, Cambridge, Mass.

Linstone, H.A. (1989) Twenty Years of TF and SC (Foreword to the Twentieth Anniversary Issue "Forecasting A New Agenda") *Technological Forecasting and Social Change*, 36 (1-2) 1-13.

Linstone, H.A. (1994) New Era - New Challenge. *Technological Forecasting and Social Change*, 47 (1) 1-20.

Linstone, H.A. and Turoff, M. (eds) (1975a) *The Delphi Method. Techniques and Applications*, Addison Wesley Publishing Company, Inc, Reading, Mass.

Linstone, H.A. and Turoff, M. (1975b) Introduction, in Linstone, H.A. and Turoff, M. (eds) *The Delphi Method. Techniques and applications*, 1-12, Addison Wesley Publishing Company, Reading, MA.

Lundvall, B.A. (1988) Innovation as an Interactive Process: From User-producer Interaction to the National System of Innovation, in Dosi, G., Freeman, C., Nelson, R., Silverberg, G. and Soete, L. (eds) *Technical Change and Economic Theory*, 349-369, Pinter Publishers, London.

Martin, B.R. (1995) Foresight in Science and Technology, *Technology Analysis and Strategic Management*, 7 (2) 139-168.

MESR, M.d.l.E.S.e.d.l.R. (1994) *Consultation Nationale sur les Grands Objectifs de la Recherche Française,* Paris, Ministère de l'Enseignement Supérieur et de la Recherche.

Meyer, L.A., Prakke, F., and Smits, M. (1993) *Foresight study of Energy Conversion Technologies* No. STB 93/10). TNO/STB.

Michael, M. (1996) *Constructing Identities,* Sage, London.

Nelson, R. and Winter, S. (1982) *An Evolutionary Theory of Economic Change,* Harvard University Press, Harvard.

Nelson, R. and Winter, S. (1977) In Search of Useful Theory of Innovation, *Research Policy,* 1 (6) 36-76.

OCV (1996) *Een Vitaal Kennissysteem: Nederlands Onderzoek in Toekomstig Perspectief,* No. OCV, Amsterdam.

Ogilvy, J. (1992) Future Studies and the Human Sciences: The Case for Normative Scenarios, *Futures Research Quarterly,* 8 (2) 5-65.

Perez, C. (1983) Structural Change and Assimilation of New Technologies in the Economic and Social Systems, *Futures* (October) 357-375.

Polanyi, K. (1983) (1944) *La Grande Transformation: Aux Origines Politiques et Economiques de Notre Temps* (Maurice Angeno Catherine Malamoud, Trans.) Gallimard, Paris.

Porter, A.L., Roper, A.T., Mason, T.W., Rossini, F.A., and Banks, J. (1991) *Forecasting and Management of Technology,* Wiley Interscience, New York.

Porter, T.M. (1995) *Trust in Numbers. The Pursuit of Objectivity in Science and Public Life,* Princeton University Press, Princeton.

Rip, A. (1995) Le Poids des Phases Initiales dans le Déroulement des Programmes, in Callon, M., Larédo, P. and Mustar, P. (eds) *La Gestion Stratégique de la Recherche et de la Technologie. L'évaluation des Programmes,* 111-124, Economica, Paris.

Sahal, D. (1985) Technological Guideposts and Innovation Avenues, *Research Policy,* 14, 61-82.

Schumpeter, J. (1934) *The Theory of Economic Development,* Harvard University Press, Harvard.

Smits, R.E.H.M. and Leyten, A.J.M. (1991) *Technology Assessment: Waakhond of Speurhond,* Vrije Universiteit Amsterdam.

Steele, B.D. (1995) An Economic Theory of Technological Products, *Technological Forecasting and Social Change,* 48, 221-242.

Twiss, B. (1994) *Forecasting for Technologists and Engineers: A Practical Guide for Better Decisions,* Peter Penegrines, London.

Van Lente, H. (1993) *Promising Technology.* PhD, University of Twente.

Van Raan, A.F.J. (ed) (1988) *Handbook of Quantitative Studies of Science and Technology,* Elsevier Academic Publishers, Amsterdam.

Vedin, B.A. (1990) *Tumbling Walls of Technology,* IMIT Metamatic, Kristianstadt.

Vinck, D. (ed). (1991) *Gestion de la Recherche.* De Boeck-Wesmael, Brussels.

Von Hippel, E. (1988) *The Sources of Innovation,* Oxford University Press, Oxford.

Williamson, O. (1996) Economics and Organization: A Primer, *California Management Review,* 38 (2) 131-145.

10   Genetics and Uncertainty

ANNEMIEK NELIS

We are, or ought to be entering an age of biomedical science and biomedical technology that could rival the magnitude and richness of the present age of physical science and physical technology. Whether we shall indeed enter this age will depend upon the attitude towards Big Biology adopted by biomedical scientists and governmental agencies that support biology. Whether the age of Big Biology will be truly rewarding will depend on the common sense and integrity of all who participate in this adventure (Alvin M.Weinberg, 1968 p.107).

Today, the biological sciences have attained the scale and capacity of 'Big Biology', not only in terms of the amount of research done in the field but also in the way in which biology promises to transform health care delivery. Weinberg, cited above, must have been a visionary man. He foresaw what was about to happen in the last part of the twentieth century. Genetic technologies for example, being one of the major successes of Big Biology, are believed to bring about a major change in the provision, delivery and organisation of health care practices. What exactly these changes will look like, however, we do not know. The future development of genetic technologies, especially genetic testing is marked by acute uncertainty.[1]

Weinberg is also right in another aspect: *what* the future will look like is not necessarily a determinant of the new knowledge or new technology. Such depends, as Weinberg acknowledges with respect to the success of Big Biology in 1964, on the activities of many different actors. Had they not put their efforts, their money and expectations behind the emergence of Big Biology, it probably would not have been as influential as it currently is. Put differently, knowledge, technology and the practices in which these are constituted, are the result of large groups of people constructing and sharing a common repertoire of truth and success. How to create a shared repertoire of rules, knowledge, habits and routines by setting priorities for the future of genetic testing, is the subject of this article.

Anticipating the Future

To set priorities for the future, as happens for example through the many worldwide foresight exercises, is not a neutral process of articulating what is 'best to be done and what should be done first'. Rather, this process reflects the ascription of expectations and promises to a particular technology or innovation area. Promises and expectations – but also threats – help shape the direction of future action and, as a consequence, help shape the direction developments might take (Van Lente, 1993). Mobilising expectations therefore is a crucial aspect of technology development for it directs actors' actions towards certain agendas in a particular direction. There is, however, no inherent outcome or an inherent 'future' embedded in technological development. Innovative technologies are the product of both decisions that were taken in the past – creating path dependencies which discourage particular options while making others more feasible – and choices that are made on the basis of a particular promise or expectation (Van Lente and Rip, 1998). If such promises and expectations are shared among a large constituency of actors, technology is most likely to stabilise. If promises and expectations are fragmented and diverse, stabilisation is less likely to occur.

A situation in which uncertainty prevails is difficult to 'manage' for the actors involved, and it is likely to prompt action which anticipates specific futures. In some contexts, but not all, this might well encourage an engagement with foresight exercises, since they are a means to set research priorities, strengthening old and new linkages between academia, government and industry whereby the nurturing of a nation's economic and social wellbeing is key. The scenarios of the future that are created within foresight exercises are not necessarily the best or the only possible future. Rather, by setting priorities for the future, technology foresight actively helps shape this future (De Laat and Larédo, 1998). Technology foresight inherently deals with uncertainties. It deals with the articulation of a common vision of the future by describing both promises and expectations through a process of bringing together a large number of actors. Foresight, one might say, is an institutionalised way to create a shared vision, idea and script for the future for actors to act upon.

Stable practices emerge as the result of the shared acceptance of certain visions, views and ideas and the material and technical products and processes in which such cognitive functions are embedded.[2] De-stabilisation might emerge when there is no longer any certainty about the knowledge used or the technologies to be applied.[3] Uncertainty means that taken for granted rules, heuristics and theories are no longer valid. However, uncertainty not only relates to things we might consider to be important in the laboratory. It might also relate to a lack of trust and consensus in the way in which new technology will be perceived or used. With respect to genetics, for example, the public image of genetic testing - the question whether people want to know what their genetic make-up entails - might disrupt the entrenchment of new technology in society.

This article explores how within the domain of DNA-diagnostic testing, uncertainties are dealt with through mobilising expectations of the future. As will be shown, the way in which the future is anticipated - either as part of formal foresight processes or not - not only depends on the uncertainties that surround the technology. It is within the wider context of innovation that we find features that both enable and constrain the capacity of institutional actors to mobilise scenarios for the future. This wider environment we will generally refer to as the *configuration* in which developments take place. Configurations refer to the pattern of relationships and interactions between the main actors involved in the development of new technologies such as genetic diagnostics (Löhnberg et al., 1999; Rappert and Brown, 2000). These interactions both shape technological expectations as well as innovation strategies.

In order to illustrate how different configurations express different ways of managing future uncertainties associated with technological development and innovation strategy, two cases have been chosen. First we will look at the development of genetic diagnostics in the Netherlands, and secondly, as a comparison, we will look into the development of genetic diagnostics in the UK. These countries are chosen because at least part of their research in this area is situated at the forefront of developments and because both countries have at their disposal a very well-regulated and well-organised genetic service. At first sight, this might suggest that genetic diagnostics enjoys a similarly stable existence in both countries. In fact, as I shall argue, genetic testing varies considerably between the two reflecting quite distinct configurational relations in each country.

Before taking a closer look at the specific configuration of the Netherlands and the UK, we will shortly discuss the particular uncertainties at stake in the field of genetic diagnostics.

Genetic Diagnostics and Uncertainty

Clinical genetics is a rather new field that, over the past twenty-five years, became established as a medical speciality. While the early days of clinical genetics were characterised by a focus on hereditary congenital disorders in small numbers of families, this has now expanded to considerably larger numbers with a more diverse range of pathologies. Genetic testing and genetic counselling - that is the provision of neutral information to families at-risk for hereditary or congenital disorders - today bear in themselves the promise of becoming one of the main drivers of modern medicine and public health. This promise is related to an important paradigm shift within clinical genetics: the shift from mono genetic to multifactorial disorders. The latter are the putative 'genetics of the future' and will outscore both in number and impact the testing for mono-genetic disorders. Mono-genetic disorders are caused by a single gene defect, the carriers of which can be sure they will develop the disease. Multifactoral disorders, however, are the result of a combination of both multiple genetic and environmental factors such as life-style, diet, life-events and so on.

Mono-genetic disorders involve rather rare disorders such as Tay-Sachs, Huntington's Disease and Cystic Fybrosis (CF). DNA-tests for multi-factorial disorders relate to more common disorders such as certain cancers, Alzheimer's disease, coronary heart disorders, diabetes and so on. It has been argued in the future we will all become 'un-patients'; healthy people with a high risk disposition towards a particular (late on-set) disease (Jonse, et al. 1996). Large numbers of individuals, often without realising so, have a more than average to very high risk of carrying a number of these multi-factorial disorders. Preventive medicine, through the deployment of genetic tests - especially at the pre-natal point - will only gain importance as the consequence of this development.[4] The shift from monogenetic to multifactoral disorders therefore is likely to foster a shift in emphasis from curative to preventive medicine (Baird, 1990; De Vries, et al., 1997).

The impact of this and related developments such as the use of DNA-technologies in forensic medicine and the potential of gene-therapy and cloning, are often presented as being the next revolution: the genetic revolution (Mager, 1991; Elmert-Dewitt, 1994). The reference to a 'revolution' points to the immense transformation and changes expected to occur as a consequence of the development and application of new DNA-technologies. Such transformation and change, however, mean that the field is increasingly characterised by indeterminacies and uncertainty. These uncertainties can be summarised as *technological, organisational* and *cultural* uncertainties.

First, with respect to *technological* developments, there are a number of issues at stake. While the early stories about DNA-testing presented a rather optimistic repertoire of 'cookbook' recipes in which every gene would be linked to a particular characteristic or trait, matters have proved to be much more complex in practice. For example, genetic researchers may use non-standard reagents and protocols in their procedures, or may produce mapping or sequencing errors. Even with respect to mono-genetic disorders which scientists thought were relatively straightforward, new mutations and complex genetic-characteristics have been described which are not always easy to explain. Also, we still know little or nothing about the role the environment (life-habits, environment, nutrition and so on) plays in triggering a particular gene to cause changes in the DNA (which might lead to disease). Research into genetic disorders is often directed towards establishing a link between genes and disorders. Such research, however, reveals little about the functioning of genes and their interaction with the environment. Obtaining knowledge about the latter process is still in its infancy.

Perhaps the most acute uncertainties arise in respect to determining *when* the actual products and benefits that follow from the application of DNA-technologies, will become available. With the prospect of the Human Genome Project, the world wide effort to map and sequence all human genes, being completed in 2000/2001, there is still considerable uncertainty about when this information will become useful for the detection, cure and prevention of disease. In general, it can be said actors do not so much doubt whether DNA-testing will be used in a broad fashion but are less sure when this will be the case.

Secondly, the shift in attention from DNA testing for monogenetic disorders to DNA-testing of multifactoral disorders is not just a technological innovation but also relates to *organisational* change. The new genetics are expected to have important consequences for the application and use of new DNA-tests and have prompted questions about how these tests should be organised and become institutionally available within health care practices.

In Europe, most countries have the facility of some form of genetic service (Harris, 1997). However, as the number of tests increase, there will be an enormous demand on existing genetic services. For the time being, it is unclear what the role will be of the major players that give shape to and are, in turn, shaped by this technology in the future. What should the pharmaceutical industry, public health agencies, biomedical companies, clinical geneticists, epidemiologists, patients-organisations and the 'public' expect and how should they anticipate the increased influence of genetic testing? It is beyond doubt these actors are highly dependent on one another for the development, introduction, monitoring and control of DNA-diagnostic tests. As will be shown below, such dependency relationships - and the historically contingent way in which they have evolved - are an important factor in the shaping of the future pattern of DNA-diagnostic tests.

Third, many of the actors mentioned above are genuinely uncertain about how the lay public in general or patient advocacy groups in particular will react to what has been coined the 'genetic revolution'. There are basic questions whether genetic diagnostics will be 'socially viable' in the long term. Sociologists increasingly point to the adverse effects of the large scale introduction of genetic testing (Nelkin and Tancredi, 1994; Spallone 1992; De Vries et al., 1997; Lippman, 1992, 1994; Hoedemacker et al., 1998; Kitcher, 1996). However, it is unclear if and how the general public will value genetic information, though it is likely that without widespread public acceptance of and engagement with diverse forms of genetic testing, the clinical use of these tests will be relatively modest.

Most of the problems and uncertainties we have so far described are broadly shared among locations where genetic testing is performed. Technical issues to do with the question when and, to a lesser extent, what genetic technologies will deliver, or how genetic services should be (re)organised and how uncertainty about social acceptability should be dealt with are concerns that are broadly shared throughout the (largely Western) world.[5] However, the organisational and institutional configurations in which these uncertainties are found and given expression, differ from context to context. The way clinical geneticists, researchers, regulators, public health agencies, patient advocacy groups and government bodies have anticipated current transformations in the field of genetic diagnostics, differs among countries as they act within different institutional and organisational configurations (Löhnberg, et al., 1999). This of course is not to say there is a simple one to one relationship between institutional configurations and the way in which problems are defined and solved. However, the way in which new configurations act as the vehicle through which problems are managed both enables and constrains future action. The history of DNA-research and clinical applications of DNA-technologies, in other words, tells us something not only about why current practices are organised as they are, but also about the way in which current developments and expectations about the future, will be accommodated.

To understand the way in which different configurations mediate the technological, organisational and cultural uncertainties associated with genetic-diagnostics, it is necessary to shed light on the *socio-historical* development of these configurations. As will be illustrated below, these configurations foster different strategies for managing expectations, in some cases ones that encourage the deployment of foresight exercises. As we shall see, foresight is not merely about the identification of technological priorities for future wealth creation, but is instead a much more sophisticated vehicle that seeks to encourage consensus on technological and social agendas. In the UK, for example, the latest foresight round on health is instructed to 'explore the potential of how the UK NHS could be used as a model for public services and healthcare systems' (OST, 1999).

A Socio-Historical Account of Genetic Testing: The Dutch Case

While the Dutch have a long history of genetic research (Meininger, 1954) it was only with the advent of the new technological possibilities created by prenatal diagnosis and chromosome-analysis in the late 1960s that genetic *services* started to emerge.

Initially the new technologies did not command a great deal of attention. However, due to the continuous efforts of a relatively small number of spokespersons, genetic counselling appeared on the agenda of the Ministry of Health. One of the most influential actors in the process of establishing clinical genetic services in the Netherlands has been the Dutch Health Council (Nelis, 1998). The Health Council (Gezondheidsraad) advises government on the state-of-the art of science and technology. However, since its advisory commissions are dominated by scientists and clinicians, the profession itself wields enormous power in determining the content of the reports thus produced (Kirejzeck, 1996). During the seventies, the Health Council published two important reports on genetic testing: *Genetic Counselling* (Gezondheidsraad, 1977), and its report on Cytogenetic Laboratories (Gezondheidsraad, 1979).

The publication of the Health Council's report Genetic Counselling retrospectively turned out to provide a blue-print for the organisation of clinical genetics. Most, if not all, of the recommendations of the committee were put into place. The committee that wrote the report consisted of both clinicians - mainly paediatricians who were involved in the provision of genetic counselling - and researchers - involved in research on prenatal diagnosis and chromosome analysis. It is perhaps not surprisingly then that this committee, in its recommendations, suggested genetic counselling should be offered only by specialised medical professionals.

Only if genetic counsellors, as the paediatricians who engaged themselves with hereditary and congenital disorders came to be known, where located at an academic hospital would they have sufficient access to resources. This environment would enable them to act in a neutral way and to be non-directive to their clients, as this is one of the major principles of genetic counselling. Genetic counsellors would obtain a special training through which they learned how to be non-directive and neutral and to make sure individuals who came for counselling would be able to make their own choices.

Also laboratory practices needed for genetic diagnostics had to be of a high quality. Postnatal chromosome analyses for example, which according to the committee often was done with out-dated methods and in laboratories which had too little experience to guarantee the quality of the diagnosis, should only be done in specialised centres. Such centres, according to the committee, could only guarantee up-to-date knowledge and methods if it had a liaison with an academic hospital.

During the early 1980s, eight centres for clinical genetics were established. Following the recommendations of the Health Council, these centres obtained a permit from the Ministry of Health which provided them with a monopoly on the provision of prenatal diagnosis,[6] chromosome analysis and, more recently, DNA-diagnostic tests. When, in the early 1980s the first DNA-test was introduced into clinical practice, it was brought under the same regulatory and organisational rules (Nelis, 1998). Within the centres for clinical genetics, genetic counselling is provided by fully trained, highly specialised medical doctors.[7]

Beside the centres for clinical genetics, a number of newly formed networks emerged. Both laboratory analyst as well as genetic counsellors established national foundations to represent their professional interests. Beside public research at universities, the Netherlands has demonstrated little commercial activity in this field. Neither the pharmaceutical industry nor the biotechnology industry has been very well developed in the Netherlands (Arthur Andersen, 1998).[8] As is the case in the development of genetic testing, industry has hardly played a role in either R&D or the provision of genetic tests. Only in the past few years have pharmaceutical companies started to take an interest in the issue of clinical genetics and genetic testing.

The configuration in which genetic testing thus developed, can be described as one in which there are quite strong, durable and interdependent relations between a relatively small range of genetics actors operating as a relatively closed network at a national level. This has meant that the technological, organisational and cultural uncertainties sketched out above have been broadly contained within this network, which has, in a sense, been 'authorised' by both the wider clinical profession and government to manage the 'new' genetics. This situation has, therefore, little to do with the size of the network or, for that matter, the size of the country, but with the way in which new technologies were taken up by a well-organised group of researchers and clinicians in co-operation with patient organisations. This group of actors forms a close network of actors with strong spokespersons to represent their interests to government, industry and the public.

Anticipating the Future

Despite uncertainty about the future of genetic testing so far, the clinical genetics network has attained what might be called a *monopoly* on the organisation, provision and diffusion of genetic technologies. There is, in other words, little room for other players in the network of clinical genetics. To illustrate this point, the Netherlands is still the only country where most genetic counselling is done by fully trained medical specialists. Others, such as non-clinical genetic counsellors - have been excluded from entry into the network.

The network is therefore more or less non-permeable, a situation that most of the actors involved would like to sustain. By emphasising the need for high quality, specially utilised centres and a limited number of locations to offer genetic counselling to as many people as required, actors not only help reproduce strong interdependencies but also a shared capacity for managing the potential uncertainties new genetic developments have provoked. The resources of the network can be deployed to cope with the demands of the 'multifactoral' world. The changing nature of the boundaries of their network can be expressed through the summary to the 1996 Health Council report:

> I would like to conclude by stating that we have always enjoyed a high level of expertise in the Netherlands where the field of clinical genetics is concerned... . It is vital that these standards be maintained and used for the benefit of more and more people with genetic risks (Gezondheidoraad, 1996, *my translation*).

Anticipating the future of genetic diagnostics does not take place in broader circles, such as for example national foresight initiatives.[9] Organising the future structure and content of genetic diagnostics happens in realms that are well-known to the actors involved in genetic diagnostics. Setting the agenda for genetic services happens through bodies such as the different professional organisations and the aforementioned National Board of Health. In the past few years, the National Board of Health produced several reports on genetic diagnostics and related subjects (Gezondheidsraad, 1980, 1989, 1994, 1998). As happened in the past, it is the spokespersons from the genetic community who participate in and chair these committees and, as such, have a strong influence on the content of these documents.[10] As was mentioned before, although the Board of Health is supposed to advise on state of the art science, it has proved to have a substantial influence on the future organisation of health care in the Netherlands (Rigter, 1992). As such, it is an important source of future co-ordination.

A Socio-Historical Account of Genetic Testing: The UK Case

Genetic counselling in the UK has a long history[11] with the first clinic having been set up as early in 1946. Despite these early experiences, the organisation of genetic services never managed to be as centralised and co-ordinated as happened in the Dutch case. Genetic counselling never fully absorbed related services such as prenatal diagnosis, hereditary metabolic disorders occurring in paediatrics and genetic research.

As happened in the Dutch case, the introduction in the early seventies of prenatal diagnosis and chromosome analysis led to an increased attention to hereditary and congenital disorders. In the UK, during this period, several reports were published which all acknowledged the need for a regionally organised Genetic Service. However, unlike the situation in the Netherlands, it appeared to be difficult to get a comprehensive national network off the ground.[12] Whereas today the UK has the facility of a regionally organised Genetics Service, it took a long time for this service to come into place. In the Dutch case, we saw a strong lobby of spokespersons leading to the publication of a very influential report that functioned as a blue-print for the organisation of clinical genetics service. In the UK, the different aspects of genetic diagnostics were dealt with in different reports and therefore remained more or less separate issues, articulated by different social actors within medical, public health, research and government.

As a result, different parts remained the responsibility of distinct professional groups who could be said to work according to their own standards and their own professional ethics. Obstetricians, for example, are still responsible for prenatal diagnosis, while paediatricians are responsible for treating many child disorders including hereditary metabolic disorders.

And since 1993, GP-fundholders can purchase clinical genetic services for relevant families.[13]

The clinical genetic services in the UK have been established mainly on an *ad hoc* basis and are said to meet only a fraction of the communities' needs. Co-operation between researchers and clinical services developed in a contingent fashion, depending strongly on the local politics of services as well as on the service correlates of new techniques.[14]

Fragmentation and, as a consequence of this, the lack of co-ordination of management and services within the NHS, has long been an issue of concern in the UK (Loveridge and Starkey, 1992; Klein, 1996). Since genetics is currently a much debated subject, renewed attention has been given to the question of how genetics should be organised both in the near and in the distant future.[15] It has generally been acknowledged that the application of genetic technologies requires an integrated approach whereby geneticists, clinicians, public health advisers, GPs and so on, work closely together.[16] Whereas a number of national and local initiatives to co-ordinate and align different services have recently been initiated, alignment as yet has not emerged. While in the Dutch case we saw a very well organised group of clinicians creating - and albeit protecting - a genetic service that would meet the criteria of high quality, specialised care and facilities, in the UK this service generally is more dispersed and fragmented.[17] Neither NHS managers, nor clinicians, have assumed the role of spokespersons for the field as a whole.[18]

Perhaps less surprising that, given the more fragmented way in which the genetics service is organised in the UK, there is much more scope for actors outside the network of medical professionals (together with patients or patient-organisations) to become part of the innovation network of genetic diagnostics. The UK has long been known for its strong pharmaceutical and biotech industry. In recent years, large pharmaceutical companies have devoted much more attention to the new genetics. While, on the one hand, this is part of the fact that gene technology is used to identify lead compounds for the production of new medicines, many companies at the same time have made a deliberate choice to turn to genetic diagnostics as one of their (future) products. The large pharmaceutical firms have positioned themselves at the forefront of developments in the field of genetic technologies (Richmond, 1999; Sykes, 1996). In general, it can be said that in light of current scientific and technological developments there has been a recent mushrooming of interest in the development of genetic technologies. Such interests are not only fed by large pharmaceutical companies but also by smaller biotechnology companies. Many of these smaller companies work on products which they hope will be marketed by big pharmaceutical firms who are able to pick up the cost of clinical trials and so on.

However, for most of the newcomers to the field of genetic diagnostics, at least some of the issues they confront are completely novel to them. As one UK interviewee from the pharmaceutical sector explained with respect to the role of public acceptability:

> ...there is a document that I saw which addressed genetic testing among other things... . It was imbued with a lot of concerns about potential risk for discrimination and marginalisation attendant on people having genetic tests. Those are relevant and important concerns but it's how do you get a link into that group? It is not the group that we normally have great contacts with.

As this extract illustrates, the firm has had to create new linkages with wider actors on the basis of little or no prior history.[19] Moreover, the 'concerns' which the company has had to address - such as risks of 'discrimination' and 'marginalisation' - pose very different political and moral issues than those that they more usually confront in Ethics Committees during clinical trials approval processes.

Anticipating the Future

When the Office of Science and Technology (OST) in the UK, in the early 1990s, organised a large foresight exercise, genetics was placed high on the agenda (OST, 1995). The UK foresight exercise, together with the Japanese Delphi exercises, has probably been one of the most centrally orchestrated foresight programmes in the world (Fleisner, et al., 1998; Rappert, 1999). The OST, which is responsible for the UK exercise, initially established fifteen panels. These were assigned the task of determining, in light of the country's strengths and weaknesses, what would be the most promising (technological) developments to invest in over the next ten to twenty years. The Panel most directly concerned with health and medicine, the Health and Life Sciences Panel, ascribed revolutionary potential to the developments within the biological sciences.

> A biological sciences revolution is under way, the impact of which will be greater than the industrial or atomic revolutions (OST 1995, p.v).

Key to the Health and Life Sciences Panel report was the acknowledgement that current and future developments in the biological sciences and in molecular biology and genetic diagnostics in particular, would have great impact on both R&D and on the provision of health care. Genetics therefore featured prominently throughout the report and was featured in four of the ten main hypotheses drawn up by the Committee.

As an institutionalised effort to deal with future uncertainties, foresight provided an opportunity for actors to guide and direct at least some of the technological and clinical developments. Although many people agreed the report contained little information they did not know before reading it, it did have an important function in legitimation and agenda-setting for research, development and clinical application. As one of the directors of a big pharmaceutical company explained:

> My diagnosis of the UK is that the impact of foresight was much more on the level of strategic thinking and the fact that all kind of organisations started to do their own foresight than the fact that they started to use the recommendations and priorities. If you look at the recommendations, you might say that is a very old fashioned list of priorities.

As we have argued elsewhere, the foresight process in the UK, both within the health care sector and private industry, has been an important source of legitimation (Rappert, 1999; Nelis and Webster, 1998). As such, it helped to put on the agenda and created awareness of, among other things, the acute uncertainties that surround genetic diagnostics.

The UK network that surrounds clinical genetics consists of a large number of actors among whom we also find relatively new players. Actors concerned and involved in genetic diagnostics include molecular biologists, clinical geneticists, paediatricians, GPs, gynaecologists, biotech companies, pharmaceutical companies, NHS-officials, public health officials and so on. Questions to do with future development and the uncertainties that surround this future - how, when and under what conditions genetic technologies will be used - are more or less dispersed across these actors. The attention drawn to genetic diagnostics through the national foresight exercise, which aims to bring together actors from industry, government and academia, fitted well into this large and openly structured network. Thus, loose dependency among a plurality of social actors has been institutionalised in the latest foresight round where different 'Task Forces' contribute on a regular basis to an open 'knowledge pool' covering

'Healthcare'. The technological, organisational and cultural uncertainties of genetic diagnostics are, in fact, more likely to flourish in such a 'policy stew'. In this sense, the framing of the three uncertainties, their articulation and the way 'answers' to them are found will move at different paces and be reflected in distinct socio-technical initiatives.

In the Netherlands, on the other hand, we found a network that from its inception has been characterised by its closeness and which has tried to keep its boundaries more or less non-permeable. What implications follow from the future development of genetic diagnostics and its related uncertainties has, so far, been mainly decided within the network itself. Uncertainty thereby has a particular locus of concern, which has primarily been within the non-permeable boundaries the particular network.

The Role of Uncertainty and Open or Closed Networks

Above I have shown how different configurations of actors deal differently with the promises and threats imposed by the development of innovative technologies. However, my aim goes beyond pointing out that different configurations articulate different answers to more or less similar questions. What I am interested in is how different configurations play out in situations where the future is saturated with uncertainty. What are the circumstances in which actors attempt to provide answers to the uncertainties they face? What strategies do they have at their disposal to do so? If, as I claim at the beginning of this chapter, the future is the product of a shared repertoire in which promises and threats become ascribed to certain parts or aspects of the technology, what does this means for instruments that are aimed at creating a shared repertoire such as foresight? It is to these questions that we will now turn.

Foresight as a Policy Tool

The common objective of foresight programmes is to improve strategic choices that will foster economic prosperity and a nation's competitive position. National foresight programmes have emerged as a fashionable means to co-ordinate science and technology policies and respond to conditions of uncertainty and change.[20] Foresight exercises, which have their roots in American and Japanese industrial policy, make use of a whole variety of methods such as Delphi studies, panel-reports, scenario studies and so on. Notwithstanding this diversity of method, it has been claimed that the priorities that are set in individual programmes do not diverge very much. So in general, there is a broad level of agreement on priority areas in individual nation states – something which is believed to be the result of the globalisation of science and technology developments (Fleisner, et al., 1998). Perhaps, a more important question therefore is not so much *what* priorities appear on the agenda but *if* something appears on the agenda at all.

I have argued earlier that uncertainty about the future of genetic diagnostics might be equally strong in the Netherlands and the UK. However, from our comparison of the configurations in which both Dutch and UK actors try to deal with the uncertainties that surround genetic diagnostics, I have suggested that dependency relationships in the Netherlands were much more structured around a closed, internally focused agenda and group than in the UK. In the UK, actors are more heterogeneous and more loosely dependent on each other. In such a situation those in the OST have been able to mobilise foresight as a vehicle through which alternative agendas and different networks might be more closely brought together. To do so implies changing the existing configuration over time, something we have yet to see. Following Van der Meulen and Löhnberg (1999) we could conclude that the answer to the question whether foresight will be used as a tool to deal with future uncertainties will depend on the nature of the configuration in which these uncertainties are to be embedded.

If, as was the case in the Netherlands, a particular configuration is very stable but also very non-permeable for outsiders, it is less likely foresight will be used as a strategy or tool to deal with future uncertainties. As was illustrated in the Dutch case, such a network might profit much more from a future-orientation that is held within the boundaries of the configuration itself. Also, because of the strong linkages within such a configuration, one might say, they are more or less capable to create their own 'local foresight exercise'. These are not formally initiated but nonetheless have a similar function as formal foresight exercises. Elsewhere, we have argued this local foresight might best be called *Future Oriented Co-ordination Activities – FOCA* (Löhnberg et al., 1999). Future Oriented Co-ordination Activity refers to the process of negotiation around expectations on developments related to one or more specific field(s) of technological innovation, with the intention to facilitate strategic positioning (ibid. p.38).

What does this mean for the future of genetic diagnostics? Although the Dutch genetic diagnostic configuration has been characterised as rather stable and closed, it is very unlikely this situation will be maintained for long. Both the interest from other actors and the increase of the number of tests, are likely to create a need for new alignments outside the configuration (see also Nelis, 1998; Rappert and Nelis, 1999). Such a process is already partly visible. As within other countries, the Netherlands installed a number of outpatient-clinics for cancer genetics. These clinics are run by geneticists in co-operation with oncologists, pathologists, gynaecologists, gasto-endocrinologists and surgeons. Also, community genetics has been placed high on the agenda of clinical geneticists. To deal with the issues raised by community genetics, alignment has been sought with epidemiologists and public health actors in particular. As a consequence of these and other developments, the network of genetic diagnostics will become more open to different actors. As for our hypotheses, this would mean that in the future there will be more space for formal foresight studies. This might be a lesson for policy makers or even the network of geneticists, to take into account. As in the UK, we may well see a much greater differentiation between the technological, organisational and cultural definitions and management of uncertainties surrounding genetic diagnostics.

Under such conditions whereby the network can be characterised as more permeable, heterogeneous and more extended than it once was, there is a greater need for fora that enable actors to assess and translate one another's interests. The routes through which this can be achieved are likely to vary considerably, but the particular emphasis on the need to resolve future questions places special emphasis on future-oriented capacity. Genetic diagnostics services are increasingly having to respond to an ever-expanding range of possible technical applications which, in turn, involves more differentiated and numerous patient groups and professional specialist constituencies.

Therefore, the longer the network and the more permeable the configuration, the greater is the requirement for future-oriented anticipatory capacity through which actors can engage with one another. As we have been able to observe, the prevalence of futures language, practice and activity is roughly proportionate to the intensity of uncertainties in these sectors. Genetic diagnostics, like other areas of innovation (such as the Internet for example), is saturated with expectations that are both relatively ill-defined, particularly in respect to multifactoral disorders, and where relationships are relatively unstable. Under such conditions, foresight may be selected as one amongst a number of appropriate sites in which to stabilise relations and set boundaries to the limits of the newly forming configurations.

Foresight panels, delphi studies and task forces are part of a process of articulating a common future, a common understanding about something that is highly uncertain. However, priorities that are set through foresight processes should not be taken at face value but seen in light of the broader context in which they emerge. If my hypothesis holds, policy makers should look for those areas where networks are unstable, heterogeneous and permeable. It is in these networks that foresight can be used to actively create a common repertoire, a shared vision of the future and, consequently, add to the stability of the network. In situations where this is not the case and networks are relatively closed and non-permeable, it might be questionable whether foresight exercises are the right tool to deal with tomorrow's uncertainties.

Notes

1. The research for this work was funded under the UK economic and Social Research Council as well as the Targeted Socio-economic Research Programme of the European Commission (SOE1-CT97-1056).
2. Rather than the power-effects of social 'normalisation' in the foucauldian sense, I am more interested in the general description of material practices.
3. Here I borrow, somewhat indirectly, on Kuhn's concept of 'paradigmatic revolution' (Kuhn, 1962).
4. See De Vries, et al., 1997.
5. See for example Harris, 1997.
6. Counselling for prenatal diagnosis of women over 36yrs however, could be done by any medical specialist or GP. Obtaining amniotic fluid would normally be done by gynaecologists. However, while being no part their remit, the laboratories for chromosome research accepted requests for diagnosis from local gynaecologists only.
7. The Netherlands in this respect is a comparatively unusual exception. Elsewhere, such as in the UK and the USA, counselling is provided both by medical doctors as well as by specially trained (genetic) nurses and social workers.
8. However, this situation is about to change. Nonetheless, the magnitude of biotech and pharmaceutical industry will by and large be small compared to other countries with a stronger history in this area (Arthur Andersen, 1998).
9. The foresight report on medicine for example, pays little attention to the subject of genetics, let alone ascribing any 'revolutionary' characteristics to this technology (Discipline Advies Geneeskunde, 1994).
10. Illustrative of this is another report, written by a philosopher on behalf of the Rathenau institute. The geneticists (researchers and clinicians) to whom the report was first presented objected to the report strongly enough to prevent its publication.
11. Among other things, the eugenic movement and interest in biological determinism has had a strong influence both on science and society (see Kevles, 1985).
12. See, for example, Ferguson-Smith (1978), Polani et al. (1980), British Paediatric Association (1979), Medical Research Council Sub-committee on Genetics (1978) and Fitzsimmons (1982).
13. However, they can not purchase the necessary laboratory tests which are funded via

District Health Authorities.
14. See also Coventry 1999.
15. See for example Zimmern (1998), the Royal College of Physicians (1996), Sykes (1996) and the Genetic Interest Group (1999).
16. Zimmern, 1998.
17. This is not to say that the Dutch network for clinical genetics has the power to actually *manage* the direction of genetic diagnostics. As I have shown elsewhere, even *within* the tight network that surrounds clinical genetics in the Netherlands, one can found a strong fragmentation. The way in which responsibilities within this network have become dispersed over a large number of different actors – including at-risk individuals who have the responsibility to decide whether they want to obtain certain information or not – makes the overall direction of genetic diagnostics unmanageable (Nelis, 1999).
18. Which of course is not to say the UK genetic service is necessarily of a poorer quality. What I want to point out here is the fact both services developed in rather different ways and, consequently, function on different basses.
19. For example, in April 1999, ten major pharmaceutical firms entered into a consortium with the Wellcome Trust to map genomic variability.
20. It must be said, however, today one can find similar uncertainties among insurers and pharmaceutical companies in the Netherlands. See for example Mirjan van Zwieten en Andre Kalden (1999) *Ons Gescreende Lichaam (our screened body),* Amsterdam, uitgeverij Balans.

References

Arthur Andersen (1998) *The Biomedical and Pharmaceutical Industry in the Northern Netherlands,* Groningen.

Baird, P.A. (1990) Genetics and health care: A paradigm shift, In *Perspectives in Biology and Medicine,* 33, 2, 203-213.

British Paediatric Association (1979) *Comments of the British Paediatric Association on the Report of the MRC/DHSS Joint Working Group on Genetic Counselling and Service Implications of Clinical Genetic Research,* British Paediatric Association, London.

Coventry, P.A. and Pickstone, J.V. (1999) From What and why did genetics emerge as a medical specialism in the 1970s in the UK? A case-history of research, policy and services in the Manchester region of the NHS, *Social Studies of Medicine,* 49, 1227-1238.

Elmert-Dewitt, P. (1994) The genetic revolution. New technology enables us to improve on nature. How far should we go? *Time,* January 17, 32-35.

Ferguson-Smith, M.A., Benson, P.F., Brock, D.J.H. et al. (1978) *The provision of services for the prenatal diagnosis of foetal abnormality in the United Kingdom,* Report of the Clinical Genetics Society Working Party on Prenatal Diagnosis in Relation to Genetic Counseling, London.

Fitzsimmons, J.A., Baraitser, B.C.C., Davison, M.A., et al. (1982) *The provision of regional genetic Services in the United Kingdom,* Report of Clinical Genetics Society Working Party on Regional Genetic Services, London.

Fleisner, P., Aguado-Monsonet, M., Bellido, F., Ducatel, K., Eder, P., Gavigan, J., Goméz Y., Paloma, S., Hernández, H., Leone, F., Moncada-Paternó-Castello, P., Rojo de la Viesca, J., Scapolo, F., Soria, A., and Vega, M., (1998) *Recent National Foresight Studies: A Rreview,* report prepared the Spanish OCYT, IPTS-JRS, Seville.

Genetic Interest Group (1999) *New Paradigms, New Policies,* Report of a seminar organised by the Genetic Interest Group, 27-29 November, Cambridge, UK.

Gezondheidsraad (1977) *AdviesIinzake Genetic Counseling,* Gravenhage: Staatsuitgeverij, Advies Gezondheidsraad.

Gezondheidsraad (1979) Advies Inzake *Cytogenetische Laboratoria* Gravenhage: Staatsuitgeverij, Advies uitgebracht door een commissie van de Gezondheidsraad.

Gezondheidsraad (1980) *AdviesIinzake Ethiek van de Erfelijkheidsadvisering (Genetic Counseling),* Gravenhage: Staatsuitgeverij, Advies uitgebracht door een commissie van de Gezondheidsraad.

Gezondheidsraad (1989) *Erfelijkheid: Wetenschap en Maatschappij, over de Mogelijkheden en Grenzen van Erfelijkheidsdiagnostiek en Gentherapie.* Gravenhage: Staatsuitgeverij, Advies uitgebracht door een commissie van de Gezondheidsraad.

Gezondheidsraad (1994) *Genetische Screening,* Gravenhage: Staatsuitgeverij Advies van een commissie van de Gezondheidsraad.

Gezondheidsraad (1998) *DNA-diagnostics,* Gravenhage: Staatsuitgeverij Advies van een commissie van de Gezondheidsraad.

Harris, R. (1997) Medical genetic services in 31 countries: An overview, *European Journal of Human Genetics*: 5, 2, 3-21.

Hoedemaekers, R. and Henk ten Have (1998) Geneticization: The cyprus paradigm, *Journal of Medicine and Philosophy,* 23, 3, 274-287.

Jonse, A.R., Durfy, S.J., Burke, W., and A.G. Motulsky (1996) The advent of the unpatients', *Nature Medicine,* 2, 6, 622-624.

Kevles, D. J. (1985) *In the Name of Eugenics Genetics and the Uses of Human Heredity,* Knopf, New York.

Kirejczyk, M. (1996) *Met Technologie Gezegend? Gender en de Omstreden Invoering van in VitroFfertilisatie in de Nederlandse Gezondheidszorg,* Utrecht, Jan van Arkel.

Kitcher, P. (1996) *The Lives to Come,* Simon and Schuster, New York.

Klein, R. et al. (1996) *Managing Scarcity,* Open University Press, Buckingham.

Kuhn, T. (1962) *The Structure of Scientific Revolutions,* Chicago, Chicago University Press.

Laat, B. de and Larédo P. (1998) Foresight for research and technology policies: from innovation studies to scenario confrontation, in R. Coombs et al. (eds.), *Technological Change and Organisation,* Edward Elgar Publishing, Cheltenham/Northhampton.

Lippman, A. (1992) Led (astray) by genetic maps: The cartography of the human genome and health care, *Social Science and Medicine,* 35, 12, 1469-1476.

Lippman, A. (1994) Prenatal genetic testing and screening: Constructing needs and reinforcing inequities, in A. Clarke (ed), *Genetic Counseling: Practice and Principles,* Routledge, London.

Löhnberg, A., Meulen, B. van der., Brown, N., Nelis, A., Rappert, B., Webster, A., Anton, F., Cabello, C., Sanz. L. (1999) *Comparative Research Design,* TSER programme, stage 2, Work Package III.

Loveridge, R. and Starkey, K. (1992) Introduction, in *Continuity and Crisis in the NHS,* Loveridge and Starkey (eds), Open University, Buckingham.

Mager, D. (1991) Whose genome project?, *Bioethics,* 4, 3, 183-211.

Medical Research Council Sub-Committee on Medical Genetics (1978) *A review of Clinical Genetics,* A Report, Medical Research Council, London.

Meiniger, J.V. (1967) *Het anthropogenetisch onderzoek in Nederland,* Geneeskundige Raad van de Koninklijke Nederlandse Akademie van Wetenschappen.

Meulen, B.J.R. van der and Löhnberg A. (in press) The use of foresight: Institutional constrains and conditions, *International Journal of Technology Management.*

Nelis, A.P. (1998) *DNA-diagnostiek in Nederland; Een Regime-analyse van de Ontwikkeling van de. Klinische Genetica en DNA-diagnostische Tests, 1970-1997,* Twente University Press, Enschede.

Nelis, A. and Webster, A. (1998) Foresight in the UK: Rhetorics into Practice, Conference Proceedings, Conference entitled *Constructing Tomorrow,* Bristol 1998, 58-71.

Nelkin, D., Tancredi, L. (1994) *Dangerous Diagnostics, The Social Power of Biological Information,* The University of Chicago Press, Chicago.

OST (1995) *Progress through Partnership: Report from the Steering Group of the Technology Foresight Programme,* HMSO, London.

Polani, P.E. (Chairman) (1980) *Counselling and Service Implications of Clinical Genetic Research.* Medical Research Council/Department of Health and Social Security Joint Working Group on Genetics, DHSS, London.

Rappert, B. (1999) Rationalising the future? Foresight in science and technology policy co-ordination, *Futures,* 31, 527-545.

Rappert, B. and Nelis, A. (1999) Anticipating the new genetics: The performance of the future, SATSU Working Paper

Rappert B. and Brown, N. (2000) Putting the future in its place, *New Genetics and Society* (in press).

Richmond, M.H (1999) *The Implications of Genetics and Genomics for Healthcare and the Pharmaceutical Industry,* School of Public Policy, UCL, January, *final draft .*

Rigter, R.B.M. (1992) *Met Raad en Daad. De geschiedenis van de Gezondheid-sraad, 1902-1985,* Erasmus Universiteit, Rotterdam.

Royal College of Physicians (1998) *Clinical Genetics Service In the 21st Century,* Royal College of Physicians, London.

Spallone, P. (1992) *Generation Games, Genetic Engineering and the Future of our Lives,* The Women's Press, London.

Sykes R.B. (1996) Foresight of advances in science and technology, in M. Peckham and R. Smith (eds.) *The Scientific Basis of Health Services,* BMJ Publishing Group, London, 1-10.

Van Lente, H. (1993) *Promising Technology The Dynamics of Expectations in Technological Development,* Delft, Enschede.

Van Lente, H., and Rip, A. (1998) The rise of membrane technology: From rhetorics to social reality, *Social Studies of Science,* 28, 221-254.

Vries, G. de, Horstman, K. and Haveman, O. (1997) *Politiek van Preventie – normatieve Aspecten van Voorspellende Geneeskunde,* Den Haag, Rathenau Instituut.

Weinberg, A.M. (1968) Scientific choice and Biomedical Science, In *Criteria for Scientific Development: Public Policy and National Goals,* E. Shils (ed), M.I.T. Press, Cambridge, MA, 107 – 118.

Zimmern, R. (1999) Genetic Medicine, Keynote speech at Genetic Interest Group Conference *New Paradigms, New Policies,* Cambridge, UK.

11 Expectations and Learning as Principles for Shaping the Future

LUIS SANZ-MENÉNDEZ AND CECILIA CABELLO

Stop visiting the past, visit the future. What a beautiful expression! Pereira said, visit the future, what a beautiful expression, I had never thought about that before (Antonio Tabucchi, *Declares Pereira*).

Introduction

We start by evoking a sentence from Declares Pereira, the novel by Antonio Tabucchi, because it expresses poetically the emerging fashion of today that encourages thinking about what the probable future may be as a principle for approaching life.

We have witnessed, over the past few years, the growing popularity of a new generation of planning tools, associated primarily with science and technology policy, but also with the management of innovation. These include among others, Foresight, technology assessment, scenario analyses, and so on. These tools aim to improve our understanding of the possible future states of the world (either as results of our own actions or others) with the objective of improving our decision making when confronting choices or selecting different courses of action. What these new approaches tell us (apart from their normative or prescriptive dimension) is that selecting a course of action or making a choice always implies the consideration of multiple futures. However, besides the normative side of any recommendations, it is also true that imaginations of the future, like imaginations of the past, are devices for living in the present (March, 1995); that is, constructing possible futures serves the present today.

More recently, some sociologists studying science and technology, whom generally consider social actions as being constrained by prevailing norms, have started to call to our attention the relevance of some variables associated with the future, such as expectations or promises. Related to this increasing interest in the future, are a variety of empirical studies published recently (e.g. Van Lente and Rip, 1998, and this volume Chapters Three and Four) that explore the role of expectations or promises in the development of science and technology. Although these empirical cases have limited ambitions in explanatory terms, there is an emerging tendency to establish generalisations based on their arguments. We would like to point to the risk of under-theorised generalisations, especially because they conflict with some empirically-based work developed elsewhere in the social sciences.

The 'future' and other associated concepts have been traditionally part of the underlying models of social science, especially in economics. In fact, what is presently the dominant model of understanding, rational choice theory, includes in its basic postulates considerations about the future states of the world as critical elements in the decision making process; there is a central concern about issues associated with the future and how possible future states of the world may influence our present choices.

Our intention in this paper[1] is not to reject these sociological or socio-economic approaches or the empirical relevance of the variables they identify, but to bring some insights drawn from behavioural decision theory and organisation theory to insist that there are other dynamics that should be considered in the process of decision making and action. We agree that the future is important for the present, but to suggest that expectations about the future are self-fulfilling is a very different matter; this is to place too much weight on the role of expectations and their impact on and between social actors.

In this chapter we interrogate those arguments about the role of expectations in science and technological development that presume a logic of rational choice, and argue for an approach which places as much - if not more - emphasis on the 'logic of appropriate behaviour' embedded in organisational culture and practice. Our aim is to show how this approach can reveal much about individual and organisational decision making and how it can contribute to our understanding of innovation strategy at the level of the firm (or indeed other organisations).

The chapter begins by briefly introducing the emergence of Foresight as a planning tool in which expectations about developments in science and technology predominate. The next section discusses recent sociological approaches to the future as a key factor in human behaviour. The following section discusses how these approaches relate to rational choice decision making models and its limitations. Finally, we will present other models that include reference to learning, experience, rules and identities which complement the role played by expectations in order to explain the behaviour of innovation actors. Our intention here is to differentiate between a model of decision-making and action based on calculative rationality and one based on highly contextualised 'reasoning'.

Fads and Fashion on the Future and Foresight

The future is what matters in the present. Although this statement may seem somewhat simplistic or obvious, what is true is the fact that forecasting, foresight, prediction and other derivatives of the future have recently become the main focus of attention concerning social, economic, and political issues, but even more so in the areas of science and technology (see Irvine and Martin, 1984 or Martin and Irvine, 1989). In particular, concerns about future developments in science and technology have led many actors, whether they be firms, organisations, or governments to engage in informal and formal foresight or future oriented processes to help determine, understand and even shape future developments in scientific and technological areas.

As a result of the efforts of improving decision-making, particularly in very uncertain environments, new tools and forms of systematic collection of information have emerged (of which foresight exercises are an extension) over the past few years throughout Europe and elsewhere (Cameron, Loveridge, et al., 1996; Gavigan and Cahill, 1997; Martin, 1994). In Europe, we have witnessed many Technology Foresight exercises, such as those in the UK, Germany, the Netherlands and France, promoted by policy makers to aid the co-ordination of science, technology and innovation policy. These exercises are now very fashionable and other countries and regions feel the need to promote and carry out Foresight programmes.

Foresight may be described normatively as a two dimensional activity: the production of information, in the form of reports, documents, etc. on future tendencies and the interaction process between actors to co-ordinate their research, development and innovation activities through a large scale consultation and mobilisation exercise. The more traditional vision is to consider Foresight as an information support device for decision making, and in this sense it has the same effect that other information tools have on the decision making process, which is primarily one of legitimation (Sanz-Menéndez et al., 2000). Foresight has emerged as part of the general dynamic of policy making linking problems and solutions, and in most cases, the purpose is to provide information in order to improve the capability of policy makers to deal with uncertainty, especially in science and technology policy.

Moreover, based on the conception of technological innovation as an inter-organisational process, Technology Foresight is also being employed by policy makers with the aim of improving the competitive position of innovation actors within national systems of innovation. Foresight activities assume that strategies and future expectations about scientific and technological development can be determined by considering medium to long term time horizons. In an explicit way the purpose of Foresight has been to align the expectations of different organisations in innovation systems, to improve the co-ordination of S&T developments in the policy context.

Furthermore, these search processes of information on futures or foresight exercises are not only occurring at national or regional levels for co-ordination and policy making purposes, but are also undertaken at the level of individual firms, research organisations and other institutions. It is undisputed that the future (and how people see the future, even in probabilistic terms) is relevant to understanding the actions of individual economic agents.

Innovation actors have expectations about future developments that shape their behaviour, and these expectations could be modelled through interaction with other innovation actors. In general, we have seen that participation in the Foresight process has brought potential benefits by accommodating mutual expectations on technological development among different innovation actors. What has attracted the attention of sociologists is how expectations matter in influencing behaviour and decision-making: however we need to build a coherent model of how they become relevant and, most importantly, how they relate to other variables that shape human action.

New Sociological Approaches to the Future and Technological Innovation

The concern for decision makers has been how to encapsulate or incorporate the future into present decisions. Policy makers have increased their demand for knowledge about the future, and more precise forecasting and predictions about the evolution of complex systems, even at the risk of changing the traditional understanding of prediction in scientific work. The use of prediction emerges as a means of legitimising policy and decision making in science policy (see, for example, Sarewitz and Pielke [1999] with illustrations from environmental policy). Within the social sciences[2] efforts have been made to find explanations about what role future science and technology may play especially in shaping developments in science and technology.

Thus, in recent years, in the social studies of science and technology, a concern about the role of expectations has emerged and has been introduced into explanatory models of science and technology development. Expectations (or other concepts related to the future) have become central concepts in arguments about developments in science, technology (S&T) and innovation.

The main sources of inspiration for these new sociological approaches are social constructivisim and socio-technical network analysis. Technological developments and changes, the adoption and diffusion of technology as well as general advancements in S&T are typically seen as the outcome of the activities of research, development and investigation processes. However these processes are embedded, constructed within, and supported by, social structures of knowledge production. In other words, technological change and evolution, with its varied rates and directions is a result of the interactions between heterogeneous actors within these social structures or networks. These interactions are governed by norms or 'rules of the game' found within these social structures. Our next step is to explore some of these core concepts.

To account for emerging technological developments, some underlying social structures have been identified: these have been described as Techno Economic Networks (TENs). The configuration and dynamics of these TENs depend on several factors including: the set of actors (the formality and durability of their relationships), network intermediaries (both human and non-human), the resource dependencies, the binding and decision rules throughout the network, the negotiation processes of issues at stake, and other processes such as the stability and irreversibility, or the dynamics of convergence and divergence of the TEN, etc. (Callon, 1986a; 1986b; 1991; 1995).

TENs are formed by heterogeneous elements and their composition is dynamic, in that they evolve over time. A TEN can thus be characterised by the degree of diversity of its composition as well as its degree of evolution. This idea encompasses concepts derived from innovation studies, which characterises innovation processes as systemic in nature rather than linear from science to the market, and that these processes involve social as well as technical aspects although they are often 'path dependent'. The overall process of building TENs is linked to 'translation', characterised by four stages: problematisation, interessement, enrolment and mobilisation (Callon, 1986a). For example, the concept of enrolment defines the manner by which actors gain audiences and join others in a common process of knowledge and technology development and utilisation. Enrolment involves aligning strategies, and coalitions can be formed through co-operation mechanisms that are used within competitive environments. However, what is of interest to us is whether, behind these coalitions mechanisms that enable the formation and mobilisation of social actors within a TEN, we can find that promises, preferences or expectations about the future play a significant role. Many authors have attempted to use these or related concepts in their empirical studies to explain processes in innovation and scientific and technological development.

Akrich (1992a) provides a good example of this approach when discussing the social construction of technology. She argues that a technical artefact can be described as a scenario which represents the roles and directions governing the interactions between actors (both human and non-human) who in turn are supposed to assume those specific roles. The innovation process must cope with the ability to manage differing relations with users whose abilities and expectations can be extremely variable. The success or failure of an expectation varies according to whether it is simply an actor's projection (their future alone) or is integrated across all socio-technical dimensions of the case. In the end, the agenda may become what Akrich (1992b) has defined as a 'script', that is, a declaration of motives, aspirations, and commitments inscribed in the expectations of the new technology. Such scripts are deployed to predetermine a future world through socio-technical prescription and impose specific actions. Along the same vein, De Laat (1996) studies the case of ADEME a French agency involved in supporting technical research in the field of energy and environment. He discusses how future socio-technical environments are implicit in the actions, the programs or technical objects developed by the agency, which in turn shape the agency's future behaviour. His contribution relates to a procedure for making explicit actors' script-based scenarios. What is implicit in his analysis is that actors use expectations embedded in the emerging technologies to define their own positions and then strategies with respect to them (see also Chapter Nine, this volume).

De Laat and Laredo (1998), exploring the relationship between innovation studies and foresight, conclude that integrating lessons from innovation studies shifts the focus of Foresight from a predictive to a procedural conception. Here they argue that the foresight processes can be seen as an arena where future scenarios of different actors meet (policy makers and researchers), thus the effects are to promote as well as confront these scenarios. The authors use two meanings of 'scenarios', first in the sense of formal scenarios used in foresight exercises, and secondly, the trajectories actors envisage which represent the world they try to construct and inscribe in their actions as attempts to stabilise or modify the TEN of which they are part.

Specific comment should be made on a significant empirical contribution by Van Lente and Rip (1998), because of its relation to future issues. It represents a practical attempt to converge two main streams in science and technology studies: the sociology of scientific knowledge and the political sociology of science. Van Lente and Rip (1998) present a case on the emergence of 'membrane technology', partially based on Van Lente (1993), as a new scientific and technological field emerging within the context of strategic science policy. But what interests us is how they describe the process of interaction between actors through cognitive and structural variables, and how that contributes to the transformation of the rhetorical space where promises become a social reality.

The basic idea in their explanation of the process of socio-technical construction relates to the effects that the dynamic of expectations has on social actors' behaviour. First a specific image or idea emerges as a direct result of the scientific entrepreneurs' actions, and around that conception a set of expectations are formed. These expectations imply that the emergent technology brings promises, which then progress as a generalised solution to problems. Expectations and promises become the tool used by research entrepreneurs and spokespersons to construct the audience, first in the policy makers world and then for firms and enterprises. Within this process, researchers, firms, and governments tend to legitimise claims of resource by promises on the development of technology. The result is a dynamic of expectations in which the rhetorical space, that is created through these claims and promises, gradually evolves into a reality that shapes the strategic actions of the actors.

Although this account is very attractive, there are some issues that need further explication, and while Van Lente and Rip derive an interesting explanatory arrangement to account for the connection between cognitive aspects and actors' behaviour, the overall argument is limited because they have forgotten almost completely the basic difference between what is anticipated and what is desired, that is, the distinction between expectations and preferences.

The authors also overlook the effects of structural interests on behaviour, the existence of conflicts of interests, while in terms of rational choice models, they fail to consider   the preferences of the actors. Consequently, a question arises: when do the actors sacrifice their preferences, because of what they expect or what they see? It is crucial to take into consideration what actors like or prefer, because it is quite plausible to imagine a situation in which besides an expectation over some technological development, actors' preferences direct their behaviour in opposite directions. The model proposed by Van Lente and Rip will not work in such a situation.

On methodological grounds it should be stated that they have selected a 'successful case' - the story of membrane technology - in which the technology becomes a social reality. They attribute the success to the expectations that are formed through 'mutual positioning'. However this mechanism appears decisive only if actors make sense of their interactions as games of strategy (Schelling, 1978). Taking some non-successful cases of technological development, such as the electric car reported by Callon (1986b), how then can we explain that Renault did not want to participate in the agenda proposed by the electricity company, even if the expectations were regarded as realisable. The idea of 'mutual positioning'   may presuppose that actors are willing to build new TENs through aligning expectations. However, as in the Renault case, this may not happen. This suggests we need to explore the notion of expectations more thoroughly. One way of doing this is by examining the arguments that economists have developed through their work on rational choice theory.

The Role of Future Expectations and Preferences in Rational Choice Decision Making Models

Economic theory, and to some extent psychology, have attributed to expectations a quite significant behavioural role based on rational choice and methodological individualism (Arrow, 1974). Economic theory has traditionally argued that human behaviour can be understood as having a large rational component even beyond the more specialised sense of maximisation, and that the dynamics of rationality largely influence choice and action. Economic analysis has even claimed that the concept of rationality has been the main 'export commodity' in its trade with other social sciences (Simon, 1978).

Some of the developments in economics have confirmed the central role of expectations in economic models. The reason is clear: virtually all economic decisions, other than the trivial, involve a consideration of the effect of time. In this context any decision must make an estimation of future possibilities. Such estimates may be based upon an extrapolation of past trends or, alternatively, may be based upon different scenarios involving optimistic or pessimistic assumptions generating a range of possible outcomes with different probabilities applied to each.

As a reaction to assumptions about traditional non-rational or naive expectations of the future, a complete line of research has been identified as 'rational expectations' theory (Lucas, 1976; Shaw, 1984). Here instead of forming expectations on the basis of limited information drawn from previous experiences, people take into account all available information. For example, when governments announce that they will do whatever is necessary to promote innovation, by taking into consideration this information on innovation policy, innovation actors would adjust their expectations accordingly.

The idea of rational expectation in economics has two components: first, that each person's behaviour can be described as the outcome of maximising an objective function subject to perceived constraints; and second, that the constraints perceived by everybody in the system are mutually consistent. The first part restricts individual behaviour as optimal according to some perceived constraints, while the second imposes consistency of those perceptions across people. In an economic system, the decision of one person form parts of the constraints upon others, so that consistency, at least implicitly, requires people to form beliefs about others' decisions, about their decision processes, and even about their beliefs (Sargent, 1993).

Although these hypotheses are quite restrictive and in some degree mechanical, economists have embraced them, applying them especially to the study of financial markets, as a reaction to a non-rational-naïve-expectations perspective. The reasoning is that if perceptions of the environment, including the perception about the behaviour of other people, were to be left unrestricted, then models of people's behaviour which depend on their perceptions could produce so many possible outcomes that they would be useless as instruments for generating predictions, and thus formal models.

Even studies on the economics of technical change have recognised for some time the important role played by expectations of future change and their influence on economic agents' behaviour. For example, Nathan Rosenberg (1976) has discussed the role of technological expectations and how these influenced strategic decisions, specifically in relation to the adoption and diffusion processes associated with innovations and new technologies. Since the technological future is obscured with uncertainty, however, different economic agents will hold different expectations and their behaviour will further differ due to varying degrees of risk aversion on the part of decision makers. The point made by Rosenberg is that overall technological expectations play an important role in the decision making processes of innovation, not only in the adoption (which may thereby be delayed) but also in determining the characteristics of the actual innovation itself.

Other economists have addressed the issue of how expectations influence a firm's decision making mechanisms. For example, Hall (1994) identifies three types of expectations. *Adaptive* expectations are those that decision makers in firms make when they revise single forecasts by correcting previous errors. *Static* expectations are those that forecast by considering that exogenous variables will remain at their current levels, while *rational* expectations are when firms form expectations according to the stochastic processes presumed in generating exogenous variables.

Underlying the basic model of understanding human behaviour in terms of rational choices there is a set of basic postulates. Rational choice is based on three assertions: universality, context representation and rationality. What this means is that every significant action is a result of choice, that to choose what course of action to take depends on the context of choice, e.g. the set of available acts and their consequences, and finally that the action chosen or selected is based on a calculation of its value (Lane et al., 1994).

Here, human action is the result of human choice, and decision making is viewed as intentional and consequential. In the most familiar form of the model, it is assumed that all alternatives, the probability distribution of the consequences they have and the subject value of each possible consequence are known; it is even assumed that the choices are made by selecting that which has the highest expected value. This emphasis on the expected value may be moderated by a risk preference (i.e. some value associated with the variability of the outcome distribution).

These underlying theories of rational choice presume a social actor has two types of expectation about the future. The first is about the future *consequences* of current actions while the second is about the decision makers' *preferences* for possible future outcomes. In the first case, choice or decision depends on the uncertain future consequences of possible current action, and although it is well recognised that human limitations may restrict the precision of the estimates, that the estimates may be biased and the information on which the estimates are based may be costly, the information about probable consequences is assumed to be decisive for the choice made. In the second case, choice depends on the preferences of individuals which are assumed to be stable, unambiguous and consistent.

Within this framework of rational choice a distinction can be made between beliefs about what a person anticipates and what a person desires, thus linking these ideas to the distinction between expectations and preferences. Desires include the ways in which actions and outcomes are defined, theories about the world are given credence, and the interpretations of those theories are elaborated. The beliefs about what a person desires or likes include affective sentiments, values, and tastes. However, the process that is postulated for coming to believe that something exists is not fundamentally different from the process for coming to believe that something is desirable, because individuals construct meaning in the context of becoming committed to the chosen action by organising arguments and information according to their beliefs (March and Olsen, 1989).

In this model, partially assumed by new sociological approaches, we can observe the important role that information plays. Sources of information, its availability, the context in which it is provided, etc. are factors which may directly influence the nature of the expectations formulated. Expectations may change in light of new information which will then influence choices, but primarily rational choice sees decisions as based on an evaluation of the consequences that different choices have for actors' preferences. This means that while expectations about a particular technology anticipate its improvement, actors may still prefer to pursue alternatives elsewhere. Here then we have a more developed notion of the relation between expectations and actions.

However, we find that there are still limitations to the simple use of this model, and its sociological derivatives. To develop our criticism of the role of and emphasis on expectations as a central concept in the explanation of behaviour we propose two further lines of argument:

a) the first one states that rational choice provides an inadequate foundation for action, in particular for understanding the innovation process in organisations. Moreover, in cognitive terms, economic agents are not the kind of entities that conceptualise their world in the ways required by rational choice. In respect to structure, economic agents interact in networks of relationships which induce processes that constrain the set of possible actions but also provide opportunities. These generative relationships are incompatible with the idea of prospective comparative evaluation of futures as defended by rational choice models (Lane, et al., 1994).

b) The second line of argument draws attention to the ways in which organisations and individuals fulfil identities and thus follow rules or procedures that they see as appropriate to the situation in which they find themselves. The logic of *consequences* (based on rationality) can be contrasted with the logic of *appropriateness*, in which actions are matched to situations by means of rules organised into identities, and neither preferences nor expectations of future consequences enter directly into the decision making process (March, 1994). This relates to the fact that individuals are not the same as organisations, especially since organisations may have conflicting objectives.

The arguments against rational choice do not pretend to imply that organisations and individuals do not make rational calculations, but rather they suggest an alternative form through which we should understand complex processes such as innovation. Our point is to insist on the danger in generalising the role of expectations as the main explanation for scientific and technological development. Our criticisms echo similar arguments made by those working in cognitive psychology and organisational studies.

From cognitive psychology and organisational theory many criticisms have been made of the factual foundations of the rational choice model, insisting on the fact that we can observe a significant bias in the choice of decision makers. For example, March and Shapira (1987) have emphasised the way in which the 'illusion of control' produces highly optimistic judgements of risks and opportunities. Kahneman and Lovallo (1994) have insisted on the adoption of 'inside views' which lead to an anchoring of plans to the most available scenarios. It is interesting that these biases in choices, usually considered as a source of failure, may be mechanisms by which the actors involved in socio-technical networks reinforce their commitment to emerging technological expectations.

Actors (individuals and organisations) make their guesses about, and use expectations as information devices within, situations to determine what to do. But the process of choosing and acting (even accepting the basic model of rational choice) depends on many other variables. The available literature has not led us to question these relationships but rather to ask if there are some other or additional factors that need to be taken into consideration.

While in the specific case of Van Lente and Rip's membrane technology the explanation could be empirically adequate to account for the process, the subsequent generalisation that expectations are the main dynamic shaping technological development is more open to doubt. Expectations are intervening variables, but cannot be considered as the main independent variable of behaviour. In logical terms, our point is that we accept that expectations are a necessary condition to explain the developments in science and technology, *but are not a sufficient condition.*

Additional Elements to Explain Human and Organisational Behaviour

We have seen that the use of expectations as a central concept in explaining human behaviour has a strong association with approaches based on rational choice models. And we have analysed some of the criticisms made about the plausibility and adequacy of those models as accounts of the real behaviour of actors. This section will develop our argument more fully, built from alternative models of human choice, on how the future influences the actions of actors.

In addition to rational calculation, actors often follow a logic that can be described in quite different terms. Instead of thinking of decisions as intended rational choices, we focus on recent studies of organisations which indicate that decisions often stem from the *logic of appropriateness* rather than the logic of consequentiality, and that decision-making processes are often better understood in terms of other consequences rather than their outcomes. It has been said that 'decisions happen' (March, 1994) instead of 'decisions are made' to suggest that the organisational process that produces decisions may be poorly understood by a simple conception of intentional future-oriented choices.

This alternative to rational, anticipatory, calculated and consequential action is based on an alternative decision logic, the logic of appropriateness, obligation, duty and rules. In fact, many of the decisions we observe reflect the routine way in which people do what they are supposed to do, and much of the behaviour in organisations is determined by standard operating procedures, professional standards, cultural norms, and institutional structures.

In this model of decision making the future is not necessarily important and the logic of human behaviour responds to norms, rules and identities that draw from past experiences or learning processes. Decisions are based on how well they 'fit' into the environment. There is no assessment of the future (no expectation formulated) and action does not necessarily form part of a choice process. While in the traditional consequential models it is implicit that choice generates action, where an expectation can be defined as something that is considered likely or certain and is used in calculations for choices. As it has been said, attributing such a central role to expectations rehearses to the traditional model of decision making processes.

The logic of appropriateness sees decision making based on rules and identities where there is a question of recognition (what kind of situation), a question of identity (what kind of person or organisation) and a question of rules (what *should* a person/organisation do in such a situation). Rule based decision making proceeds differently from rational decision making because its establishes identities and matches rules to recognised situations (March 1994). Rather than evaluating alternatives in terms of the values of their consequences or adjusting to the emerging expectations, rules of appropriateness match situations and identities. For example: in terms of situation, how do I define what kind of a situation I am confronting? This might be a situation in which the research actor has discovered some property or feature in the laboratory. In terms of identity , we can ask what kind of position or status do I have - as R&D manager, as researcher and so on. In terms of matching, what is appropriate for a person like me in a situation such as this? In this case, before strategic science emerged one would probably have been only concerned with the publication of the results, but in the last two decades new rules have emerged to increasingly prescribe that (academic) research should try to disseminate and mobilise its findings to interested 'users'. If we revisit the case of membrane technology, even though expectations are relevant in the explanation of the behaviour of researchers, we could also consider the changes in identities of researchers and especially the emergence of new rules as a result of a new strategic science.

The rule following behaviour is not wilful in the normal sense. It does not stem from the simple pursuit of interest and the calculation of the future consequences of current choices. That is neither preferences nor expectations play the key roles; rather, it comes from matching a changing (and often ambiguous) set of contingent rules to a changing (and often ambiguous) set of situations. Rule following can be viewed as an implicit agreement to act appropriately. The existence and persistence, the development and transformation of the rules is then the basic issue for the explanation. The understanding of what is appropriate evolves over time, and current rules store information generated by previous experiences. Thus roles may be seen as coded information.

However, looking at decision and action we can not forget in our analysis the environment (of organisations and individuals), that is the conditions for adaptive reactions, because many specific changes in organisations have resulted from the desire to survive in particular competitive environments.

Rules and their environments adapt to each other by means of several intertwined processes. These processes by which identities and rules come to anticipate the future or reflect the past include two main aspects according to March (1994): *Analysis*, that involves anticipation and evaluation of the future consequences by intentional decision makers - a forward looking process. These are theories based on analysis as the primary mechanism of adaptation presume that rules reflect expectations of the future; *Bargaining* which is a process of negotiation, conflict and compromise among decision makers with inconsistent preferences and identities - that can be either forward or backward looking process or both.

Furthermore, there are three major processes by which rules develop: *Imitation* which involves the copying of rules, practices and forms used by others, which is either a forward or backward looking process or both. Imitation, is a common feature of organisational adaptation, and decision making can be seen as reflecting rules that spread through a group of organisations like fads or fashions. *Selection* considers the differential birth and survival rates of unchanging rules and decision making units that use them, it presumes that rules reflect history. Selection, is a process that could be identified if the individual rules are invariant, but the population of rules changes over time through differential survival. Finally, *learning* comes from experience-based changes of routines and from the ways routines are used. It also presumes that rules reflect history. Learning is a process by which actors or organisations modify the rules for action incrementally as a result of feedback from the environment. Such experiential learning is often adaptively rational.

Theories of rational action in decision making processes presume that expectations and wilful actions of human beings enact the future in the present. Rational actor models explain adaptation in organisational rules and forms as a result of the preferences of actors and their calculations of future consequences. In contrast, theories of identities, rules and institutions tend to emphasise history dependent adaptation for the decision making process. The *past* is seen as imposing itself on the present through retention of experience in routines. Historical processes by which the present encapsulates the past are the mechanisms of theories of change, including theories of learning, culture and natural selection[3] (March, 1995).

However the adaptiveness of organisations to their environment and uncertainty involves both the *exploitation* of what is known and *exploration* of what may come to be known (Levinthal and March, 1993; March, 1991). Exploitation refers to short term improvement, refinement, routinising and elaboration of existing ideas, paradigms, technologies, strategies and knowledge. It emphasises improvement of existing capabilities, competencies and technologies. Meanwhile, exploration involves risk taking and refers to experimentation with new ideas, paradigms, technologies, strategies and knowledge in the hope of finding alternatives that improve on old ones.

Exploitation and exploration are both necessary for organisations, because exploration produces variety in experience (experimentation, variation, diversity) while exploitation produces reliability in experience (selection, consistency, unity) (March 1994, 1995). The problem is one of finding the balance between the two because survival depends on both. Rational choice theories represent that balance as a problem of balancing search and action; institutional change models the problem of balancing change and stability; and theories of evolution see it as the problem of balancing variation and selection.

But the dynamics of learning tends to destroy the balance, because the returns of exploitation of existing knowledge are systematically closer in time and space than are the returns of the exploitations of possible new knowledge emerging from exploration. This situation produces two traps for adaptive systems: the 'failure' trap and the 'success' trap (Levinthal and March, 1993; March, 1991). In this context, the imagination of possible futures is a mechanism used by organisations in helping them in their process of exploration and experimentation because the proposed futures insulate exploratory ideas from the hostile environment. Inventing the future, either from the past or from future imagination, serves to stabilise organisational understanding and expectations. However, while imagining futures serves the process of exploration, the consequences of insisting on the future *and forgetting the present* (that is with regard to capabilities and competencies) would be disastrous for individual organisations.

Again we have to remember that choice and action are not interchangeable terms. In complex foresight horizons (Lane and Maxfield, 1995) characterised by rapid change, uncertainty and ambiguity, firms' survival should consist of an on-going set of practices that interpret and construct the relationships that comprise the world in which they act. Interpretation means making sense of what is happening and to act on the basis of this understanding. In this sense, we find that the most important actions that agents can take are those that enhance the generative potential of the relationships into which they enter with others. As a result, they must learn to set aside prior expectations and plans and follow where the relationships lead. Interaction is a mechanism providing vital information (either by transforming expectations or learning from others) and thus influencing the behaviour of actors in contexts of mutual dependence.

Nevertheless, although individuals and organisations follow rules and identities, this is not to say that their behaviour is always predictable. Rule based behaviour contains uncertainty, while situations, identities and rules are often ambiguous. Decisions (and actions) depend on processes of recognition that classify situations, processes of self-awareness that clarify identities, and processes of search and recall that match appropriate rules to situations and identities. All these processes are reasoned action but they are different from the processes of rational analysis (March, 1994).

Conclusion

Recent studies have brought to our attention the relevance of some variables associated with the future, such as expectations or promises. We would argue that expectations about the development of science and technology emerge and are constructed within socio-technical networks known as TENs. The concept of TENs is based on the idea that innovation processes are non-linear and systemic and involve social as well as technical aspects, although they are often found to be path dependent.

This chapter has attempted to question the relationship between expectations and decision making processes. In particular we must remember that decisions in organisations, involve a complex ecology of factors: 1) trying to act rationally with limited knowledge (and expectations) and preferences coherence; 2) trying to discover and execute proper behaviour in ambiguous situations; and 3) trying to discover, construct, and communicate interpretations of a confusing world. However, in this complex situation, to characterise the action process of actors only in terms of expectations would be to misrepresent what is actually involved in innovation within organisations.

Expectations may be considered a mechanism of relevance when talking about individual action and its consequences. When we address the problem of organisations (firms, etc.) and their expectations, what we confront is a more complex problem of the construction of social expectation. It is inappropriate to apply simple explanatory models developed for individual behaviour to account for the behaviour of organisations and collective actors.

Expectations have played a relevant empirical role in various cases - such as that of membrane technology, but the general foundations of the behaviour of the actors and the outcomes of the process of technological development have been neglected. Only within rational choice models do we find a coherent theory of expectations. However expectations, as future images, incorporate only part of the elements to be considered in action or choice for the decision maker. Learning, rules and identities also play a central role.

Finally, we have to distinguish between what expectations are about. There is a difference between expectations about future technological development and expectations of how other actors will behave. From our position, the central point is not the formation of expectations about the paths, timing or developments of a new technology, since once they are recognised they may just become parameterised. What is important are the expectations of the behaviour of other actors, because it is with respect to that behaviour that the decision maker should respond. In many cases the factor is not how technology will evolve, since uncertainties are well known, because of path dependencies, lock-in, etc., but rather how other actors will respond to this given or even shared expectation. There may also be a dominant expectation within a TEN caused by a leadership coalition or social power relationship among the actors. In these contexts what counts is the interaction between actors where generative relationships emerge.

While a complete description of the overall process is probably elusive, and the limitations are evident, we have tried to point to the necessity of locating expectations and future images into a broader context for understanding the dynamics of actors and organisation decision making and technological development.

We do not, of course, reject the significant role futures and expectations play in human and organisational behaviour. Expectations and images of the future help life to be less trying. As Eva Luna, from Isabel Allende, recalled in the imagination of her mother, Consuelo:

> She manufactured the substance of her own dreams, and from those materials constructed a world for me... to make our journey through life less trying (Isabel Allende, *Eva Luna*).

Notes

1.   This paper has been produced in the context of the research project Foresight as a tool for the Management of Knowledge Flows and Innovation, funded by the IV R&D

Framework Programme of the European Union (TSER Programme -SOE1-CT97-1056) and the III National R&D Plan (SEC-98-1539-CE). The authors acknowledge the comments, criticisms and editing from Andrew Webster. Usual disclaimers apply.

2. We should also mention that in other research areas such as comparative policy or public policy we have also witnessed the emergence of ideas or beliefs (Hall, 1989, 1993; Goldstein, 1993; Goldstein and Kehoane, 1993) and epistemic communities (Haas, 1992) as independent variables explaining sharp policy turns in many policy domains; a revival of the traditional arguments from Max Weber that ideas (worldview, causal models, etc.) are a decisive variable -in association to interests- to understand human behaviour.

3. We should state that there is convergence between our organisational arguments and some of the new economic arguments that convey relevance to the trajectories of development and the irreversibility of some dynamics and processes, such those stated by Brian Arthur (1989; 1990) and Paul David (1985).

References

Akrich, M. (1992a) Beyond social construction of technology: the shaping of people and things in the innovation process, in Dierkes, M. and Hoffmann, U. (eds), *New Technology at the Outset,* Campus Verlag, Frankfort/ New York, 173-191.

Akrich, M. (1992b) The De-Scription of Technical Objects in Bijker, W.E. and Law, J. (eds), *Shaping Technology/Building Society: studies in sociotechnical change,* MIT Press, Cambridge Mass, 205-224.

Allende, I. (1989) *Eva Luna,* Bantham Books, London.

Arrow, K. J. (1974) Limited Knowledge and Economic Action, *American Economic Review,* 64, 1, March, 1-10.

Arthur, B. W. (1989) Competing Technologies, Increasing Returns, and Lock-in by Historical Events, *The Economic Journal,* 99, 394, March, 116-131.

Arthur, B. W. (1990) Positive Feedbacks in the Economy, *Scientific American*, February, 92-99.

Callon, M. (1986a) Some elements of a sociology of translation: domestication of the scallops and the fishermen of St Brieuc Bay, in Law, J. (ed), Power, Action and Belief. A New Sociology of Knowledge *(Sociological Review Monograph* 32), Routledge and Kegan Paul, London, 196-233.

Callon, M. (1986b) The Sociology of an Actor-network: The Case of the Electric Vehicle, in Callon, M., Law, J. and Rip, A. (eds), *Mapping the Dynamics of S&T*, Macmillan, Houndmills, 19-34.

Callon, M. (1991) Techno-economic networks and irreversibility, in Law, J. (ed), *A Sociology of Monsters,* Routledge and Kegan Paul, London, 132-161.

Callon, M. (1995) Technological conception and adoption network: lessons for the CTA practitioner, in Rip, A., Misa T.J. and Schot J (eds), *Managing Technology in Society,* Pinter Publishers, London, 307-330.

David, P. (1985) Clio and the Economics of QWERTY, *American Economic Review Proceedings,* 75, 332-337.

De Laat, B. (1996) Scripts for the Future. Technology foresight, strategic evaluation and socio-technical networks: the confrontation of script-based scenarios, PhD thesis, University of Amsterdam.

De Laat, B. and Laredo, P. (1998) Foresight for research and technology policies: from innovation studies to scenario confrontation, in Coombs, R., Green, K., Richards, A. and Walsh, V. (eds), *Technological Change and Organization*, Edward Elgar Publishing, Cheltenham/Northampton, 50-179.

EC (Report by Cameron, H., Loveridge, D. et al (1996) *Technology Foresight: Perspectives for European and International Co-operation*, Brussels-Manchester, EC/DGXII-PREST.

Gavigan, J.G. and Cahill, E. (1997) *Overview of Recent European and non-European National Technology Foresight Studies*, IPTS, EC-JRC, Seville.

Goldstein, J. (1993) *Ideas, interests, and American trade policy*, Cornell University Press, Ithaca-London.

Goldstein, J. and Kehoane, R.O. (1993) Ideas and Foreign Policy: An Analytical Framework, in Goldstein, J. and Kehoane, R.O. (eds), *Ideas and Foreign policy. Beliefs, institutions, and political change*, Cornell University Press, Ithaca-London, 3-30.

Haas, P.M. (1992) Introduction: epistemic communities and international policy coordination, *International Organization*, 46, 1, Winter, 1-35.992.

Hall, P. (1994), *Innovation, Economics and Evolution*, Harvester Wheatsheaf, New York.

Hall, P. A., (ed) (1989) *The Political Power of Economic Ideas. Keynesianism across Nations*, Princeton University Press, Princeton (NJ).

Hall, P.A. (1993), Policy paradigms, social learning, and the State, *Comparative Politics*, 25, 3, April, 275-296.

Irvine, J. and Martin, B. (1984), *Foresight in Science: Picking the Winners*, Pinter Publishers, London.

Kahneman, D. and Lovallo, D. (1994) Timid Choices and Bold Forecast: A Cognitive Perspective on Risk Taking, Rumelt, R.P., Shendel, D.E. and Teece, D. (eds), *Fundamental Issues in Strategy: A research Agenda*. Boston: Harvard Business School Press, 71-96.

Lane, D. and Maxfield, R. (1995) *Foresight, Complexity and Strategy*, SFI Working Paper 95-12-106.

Lane, D., Malerba, F., Maxfield, R. and Orsenigo, L. (1994) *Choice and Action*, SFI Working Paper 95-01-004.

Levinthal, D.A. and March, J.G. (1993) The Myopia of Learning, *Strategic Management Journal*, 14, 95-112.

Lucas, R.E. (1976) Econometric Policy Evaluation: A Critique, *Journal of Monetary Economics*, 2, Supplement.

March, J.G. (1991) Exploration and Exploitation in Organizational Learning, *Organizational Science*, 2, 71-87.

March, J.G. (1994) *A Primer on Decision Making: How Decisions Happen*, The Free Press, New York.

March, J.G. (1995) The Future, Disposable Organizations and the Rigidities of Imagination, *Organization*, 2 (3/4), 427-440.

March, J.G. and Olsen, J.P. (1989) *Rediscovering Institutions: The Organizational Basis of Politics*, New York, The Free Press.

March, J. G. and Shapira, Z. (1987) Managerial perspectives on Risk and Risk taking, *Management Science*, 33, 1404-18.

Martin B. (1994) Technology Foresight: A Review of recent government exercises, included in the special issue of the OECD journal *STI Review*, 17, 1996 on Government Technology Foresight Exercises, 15-50.

Martin, B. R. and Irvine, J. (1989) *Research Foresight: Priority-Setting in Science,* Pinter Publishers, London-New York.

Rosenberg, N. (1976) On technological expectations, *Economic Journal,* 86, September, 523-535.

Sanz-Menéndez, L., Cabello, C. and Garcia, C. E. (2000) Understanding Technology Foresight: the relevance of its S&T policy context, *International Journal of Technology Management* Special issue on Technology Foresight, (forthcoming).

Sarewitz, D. and Pielke Jr., R. (1999) Prediction in science and policy, *Technology In Society,* 21, 121-133.

Sargent, T. J. (1993) *Bounded Rationality in Macroeconomics,* Clarendon Press, Oxford.

Schelling, T. C. (1978) *Micromotives and Macrobehavior,* W.W.Norton, New York.

Shaw, G. K. (1984) *Rational Expectations. An elementary exposition,* St. Martin's Press, New York.

Simon, H. A. (1978) Rationality as Process and as Product of Thought, *American Economic Review,* 68, 2, May, 1-16.

Tabucchi, A. (1996) *Declares Pereira,* New Directions, New York.

Van Lente, H. (1993) Promising Technology: the dynamics of expectations in technological developments, PhD thesis, University of Twente.

Van Lente, H. and Rip, A. (1998) The rise of membrane technology: from rhetorics to social reality, *Social Studies of Science,* 28, 2, 221-254.

12   Contested Health Futures

TOM LING

Introduction

Five centuries ago, Hieronymus Bosch painted his Millennium Triptych. As John Berger wrote, this vision of Hell presents us with a boundary-less future:

> There is no horizon there. There is no continuity between actions, there are no pauses, no paths, no pattern, no past and no future. There is only the clamour of the disparate, fragmentary present. Everywhere there are surprises and sensations, yet nowhere is there any outcome. Nothing flows through: everything interrupts. There is a kind of spatial delirium (Berger, 1999).

Cambridge fans will recognise this as a concise version of their team's tactics on a Saturday afternoon but Berger is making a more serious point. He sees this vision of hell in aspects of today's society, in the news coverage of CNN, for example, and throughout the world's mass media. This vision cannot provide the basis for purposive, consensual action.

This sense of boundary-less time is more evident in some aspects of society than others. Virtual culture, in particular, offers us the opportunity to arrange events not according to their chronological or spatial ordering but according to their context and use. Castells shares Berger's dystopic vision of these aspects of society.

> Thus it is a culture at the same time of the eternal and of the ephemeral. It is eternal because it reaches back and forth to the whole sequence of cultural expressions. It is ephemeral because each arrangement, each specific sequencing, depends on the context and purpose under which any given cultural construct is solicited. We are not in a culture of circularity, but in a universe of undifferentiated temporality of cultural expressions (Castells, 1996).

However, chronological time still operates in most social settings even if challenged by the unpunctuated 24 hour day, the seven day week, and flexi-working. This challenging the edges of chronological time may erode still further a belief that chronology gives us progress but it does not remove the sense that there will be a future; rather, it heightens our anxiety about what that future might be.

In health policy as elsewhere in society, therefore, Bosch's garden of lost souls is not the only source of future visions. Alternative visions are actively created in various settings. These must contend not only with Berger's and Castells' concerns but must also address a growing sense of unease from within the healthcare system about an apparently unpredictable and threatening future. These efforts to establish and stabilise competing accounts of the future can be found in White Papers and ministerial statements, in the health task force of the UK Office of Science and Technology's Foresight Programme, in the regional and local plans of the Department of Health and in countless interactions on the wards, workshops and offices of the NHS.

The Futures of Healthcare

The arena we are examining - the British NHS - is unique in being a centrally funded and directed healthcare system; its size and complexity make it a fascinating case study for anyone interested in organisational dynamics in the late modern world. In common with all western healthcare systems it is the product of an attempted progressive exclusion of folk medicine, witchcraft and fraudsters from treating the sick. In its place has emerged a highly complex set of institutions and activities which require some form of alignment. For the first fifty years of the history of the NHS, accounts of the future were generated by trusted professional, bureaucratic and political processes. However, we will see how, in the face of 'the clamour of the disparate, fragmentary present', 'the future' no longer holds the healthcare system together. In fact it increasingly threatens to fragment it. It is in this light that we should view the increasing attempts to provide tools which stabilise shifting future visions. We will see that this is an unavoidable exercise for healthcare organisations but we will also see that these activities privilege some interests and values while suppressing others.

Until very recently, visions of the future from within the British healthcare sector accurately mirrored the wider disposition of power within its institutions. This situation has been summarised by Harrison, et al., (1990) as having nine characteristics. These might be paraphrased as:

- Health policy change was incrementalist because the centre had to negotiate change with the periphery.
- No individual agent could impose change; change was achieved through partisan mutual adjustment.
- Within this process, the medical profession had a significant veto power.
- Lay interests were relatively weak.
- Consumer interests were weak and management introverted.
- The core was able to exert some control over the global sum allocated to the NHS.
- Management in healthcare was reactive, consensus-seeking and conflict-averse.
- The distribution of power could often lead to policy inertia.
- The essential architecture of the NHS was underpinned by a long-lasting political consensus.

In this context, the future was viewed in a particular way. Essentially, the future was plannable, negotiated and incremental. Turbulence from outside the NHS, in the shape of changing popular values, new technologies, demographics and epidemiology could all be managed through the partisan mutual adjustment of the key players. Academic commentators broadly agreed with this characterisation and argued only over whether rational choice theory, corporatism, elitism or pluralism provided the best model for explaining what happened.

However, in a textual analysis of the health White Papers of 1972, 1989 and 1997, Ian Greener shows how the discourse of change shifts over time (Greener, 1999). In the 1970s, the emphasis was heavily upon the fact that the changes being proposed had been widely agreed on, and that they were a rational adjustment to emerging circumstances. By 1989, the White Paper started off reaffirming the government's commitment to the values of the NHS but it then quickly emphasised that rising demand and new medical technologies were forcing a change in the delivery of healthcare. Here the future begins to appear as something which forces us to act today - irrespective of what key players in the NHS want. By 1997, the White Paper began by suggesting that the NHS had unacceptable waiting lists and variations in quality, and too much bureaucracy. A vision of a 'new' NHS was offered; one of partnership, integrated care, greater responsiveness, and improved quality. Above all, the urgency of the situation and the unavoidability of change was stressed (although change should not be revolutionary and should stretch beyond the lifetime of one government).

This new sense of change is not only found on the pages of White Papers. According to the editor of the British Medical Journal (BMJ), the Journal is now constantly needing to address the future whereas in the

1970s the word 'future' itself would hardly have appeared in this cultural barometer of the British medical profession.[1] The November 13th volume of the BMJ (no 7220) is entitled 'The Impact of New Technologies in Medicine' and its editorial asserts 'for most of medicine the future is highly uncertain'. Its concluding remark 'don't be scared of the future' feels a little like the conversation between the two heroes towards the end of the film *Butch Cassidy and the Sundance Kid* just before they burst out of a doorway and into a very short and terminal future, frozen for ever in the final frame.

The sense that change is not what it used to be is also captured by Rudolf Klein, a respected commentator on the British NHS, in his final chapter for the third edition of *The New Politics of the NHS*. This chapter is called 'Ambiguous past, uncertain future'. In it Klein comments:

> On one point there appears to be total unanimity about the future of health care. This is not only that the future will be different from the past, as always, but that it will be so in ways that are extraordinarily difficult to predict (Klein, 1995 p.247).

This sense of uncertainty combined with a sense of the acceleration of change is found throughout many of the futures-oriented projects in and around the NHS (see for example, the 'Madingley Scenarios' in Ling, 1999a).

Uncertain Futures: The Cases of Genetics and Information Technology

Throughout much of the literature on the healthcare system in Britain there is an assumption that the system faces a 'paradigmatic shift' or a 'revolution'. Driving this, it is often claimed, are new technologies, including new pharmaceutical products and bioinformatics (another non-technological driver is often said to be demographic change). In this section we look briefly at the claims made around two of these innovations; human genetics and information technologies.

Throughout the medical and public policy-oriented literature there is a widespread assumption that the new human genetics will radically transform the future. This claim is related to at least one of three issues. The first is that its application will transform healthcare. The second is that, as biology replaces physics as the driving force in science there will be a new scientific and technological revolution. Thirdly, it is claimed that the new genetics will transform our relationships to each other and to our selves; it will create the 'geneticisation' of culture with a new iconography of the gene. We focus here mainly on the healthcare aspects of this literature.

Walter Gilbert claimed in 1992 that 'the possession of a genetic map and the DNA sequence of a human being will transform medicine' (Kevles and Hood, 1992). The Nuffield Council on Bioethics' Report *Genetic Screening: Ethical Issues* in 1993 argues that the astonishing speed, inescapable effects, and the fear of interfering with the basis of life itself all go to make genetic research different. Rothman (1998) even suggests that genetic engineering might have as profound an effect of society as all previous technologies put together.

It is also widely acknowledged that there is a potential 'dark side' to the new genetics. Tom Wilkie (1993) talks of 'perilous knowledge' and Appleyard (1999) subtitles his book 'staying human in the genetics future'. Marinker also asks how we will 'stay human' in the genetic age and suggests 'As *Homo sapiens* prepares to depart, *Designer Man* may already have made her tentative appearance within the half-century ahead' (Marinker, 1998). What this slouching beast will be like we cannot tell, but it will be the product of decisions taken today with unknowable implications for the future.

A second set of claims around how healthcare will be transformed by a new technology relates to information and communication technologies. This is often linked to the claim that we are moving towards a new 'information society'. The key components of this trend are well known:

- new information technologies each able to communicate with the others;
- new ways of 'talking' to computers (touch-screen, voice, etc.);
- robotics - computerised aids to human work;
- globalisation of communications leading to the global village for IT users.

The Information Society has been defined as:

> A society characterised by a high level of information intensity in the everyday life of most citizens, in most organisations and workplaces; by the use of common or compatible technology for a wide range of personal, social, educational and business activities; and by the ability to transmit and receive digital data rapidly between places irrespective of distance (The INSINC Working Party definition of the information society, in The Net Result, IBM, 1997 p.5).

Some even see this as providing the basis for a fundamental shift in healthcare. As a BMJ editorial has argued, information age technology could be used to focus resources far more effectively on the individual carer and harness the capacity for self-care.

> In 'industrial age medicine' medical professionals hardly recognise the large amounts of self care that occurs but under 'information age medicine' 'professional care will be viewed as the support to a system that emphasises self care. Healthcare managers will progress in this world from managing disease to promoting health, and they will do this though lifetime plans that are built on intimate and detailed knowledge of customers' (BMJ, 1997).

Whether through liberating the capacity for self care or through other means, there is a widespread view that the age of health informatics will provide significant added value to medical encounters (Marinker and Peckham, 1998). In summary, the implications for the healthcare system are said to be that:

- the development of telemedicine will lead to the centralisation of radiological images and many diagnostics such as skin lesions, and greater provision of home care;
- telemedicine will give more people easier access to greater expertise;
- patient care and advice will require easy access to a work-station;
- geographical distance will become less of a problem (and this will also be associated with transnational and global developments);
- health informatics will make a more unified system of health care more achievable;
- there will be the development of a more informed consumerism.

To this should be added the role of information technology in monitoring outcomes, measuring performance, ensuring quality, managing risk, and spreading best practice. Coote and Hunter (1996) broadly agree that 'bio-genetics, information technologies and interactive communications undoubtedly have the capacity to transform the NHS beyond recognition within a few years'. They also point out that 'organisational and professional interests can combine to produce formidable resistance to any kind of change' (the implication being that the impact of these technologies is only resisted by sectional interests).

Although many commentators stress the impact of one or another technology, others also stress that it is the inter-locking nature of different new technologies which matters. For example:

> The 21st century will be marked by the completion of a global information superhighway, and by mobile telecommunication and computing power,

thus decentralising and diffusing the power of information, delivering the promise of multimedia, and enhancing the joy of interactive communication. In addition, it will be the century of the full flowering of the genetic revolution, For the first time, our species will penetrate the secrets of life, and will be able to perform substantial manipulations of living matter. While this will trigger a dramatic debate on the social and environmental consequences of this capacity, the possibilities open to us are truly extraordinary (Castells, 1997).

However, the possibilities may be extraordinary but as information and biotechnologies combine in unpredictable ways we cannot begin to chart the consequences for social life and environmental degradation. We have no way of knowing whether the new genetics will lead to a massive step forward in human health and happiness or whether, on the contrary, it will produce eugenics, transgenic diseases, a catastrophic loss of biodiversity or what not. We cannot know what opportunities may arise for web-based crime or for an intrusive state made possible by information technologies intended to improve the lot of human beings. The reality is that in this world, knowledge, and not ignorance, becomes a source of danger: 'not a deficient but a perfected mastery over nature' (Beck, 1992). This creates a new dynamic of risk which requires each individual to respond to the new knowledge about uncertainties. It becomes unacceptable for individuals to refuse to respond to the known risks of their behaviour. This transforms the public-private divide and becomes the 'motor' of self-politicisation (ibid.). New risks are therefore constructed, which require new forms of governance and governmentality to manage (see Ling, forthcoming).

Responding to Uncertainty

Uncertainty faces the healthcare system on all sides. We have briefly noted the new genetics and information technology. Others emphasise other new medical technologies, the rise of consumerism, the impact of globalisation, the erosion of trust in professionals, and changing ethics. This is a heady mixture of uncertain futures which calls for a response.

One response to this growing uncertainty is to seek to dispense with some earlier organisational forms and work-place cultures. An early target was 'bureaucracy'. The influential text by Osborne and Gaebler (1992) might stand for a whole range of ideological attacks on bureaucracy from within and without the healthcare sector. They argue that in an uncertain environment, both private and public bureaucracies will fail. In a rhetoric which echoes through the 1990s, they go on to argue:

Today's environment demands institutions that are extremely flexible and adaptable. It demands institutions that deliver high-quality goods and services, squeezing ever more bang out of every buck. It demands institutions that are responsive to their customers, offering choices of non-standardized services; that lead by persuasion and incentives rather than commands; that give their employees a sense of meaning and control, even ownership. It demands institutions that empower citizens rather than simply serving them (Osborne and Gaebler, 1992).

Note that the uncertainty about the environment does not prevent us from knowing what 'it' demands: flexibility, a sense of ownership, responsiveness and empowerment. Future uncertainty is used as a stick to beat the bureaucrats and to promote an alternative vision. This sleight of hand is often repeated. For example, the introduction to the discussion document *A First Class Service* (Department of Health, 1998) notes early on that the NHS faces major challenges:

They are challenges that are common to other healthcare systems elsewhere in the world, coping with greater and faster medical advances. The challenges posed by a better informed and more demanding public. And the challenges that come from shifts in family structures, changes in working life and an ageing population (Department of Health, 1998 p.5).

The document uses an imagined future as a basis for encouraging new ways of working; partnerships, indicators and national standards. In the late 1980s the same imagined future was used as a basis for encouraging 'entrepreneurial governance', for the wider use of contracts as a basis for organising inter-organisational relations, and for new forms of human resource management (see du Gay, 1996; Ling, 1994).

If the first response to uncertainty was to de-bureaucratise, the second was to advocate cultural change. 'Governing by culture' focuses on getting individuals and organisations to think and act differently without the need for formal organisational change. Therefore new incentives and penalties are created. For individuals these might range from new social security regulations for lone parents to the creation of healthy living centres. It is entirely consistent with the need to use risk as a means towards self-politicisation that the White Paper on public health *Our Healthier Nation* (CM 4386, 1999) should include a message from the Chief Medical Officer. White Papers used to be documents laying out, in technical language, the objectives of the government and how these would be achieved. By 1999, however, the audience is no longer only the policy technicians and politicians. It now includes all citizens who are being given new responsibilities in meeting health targets. The Chief Medical Officer's 'Ten Tips for health' are:

- Don't smoke. If you can, stop. If you can't, cut down.
- Follow a balanced diet with plenty of fruit and vegetables.
- Keep physically active.
- Manage stress by, for example, talking things through and making time to relax.
- If you drink alcohol, do so in moderation.
- Cover up in the sun, and protect children from sunburn.
- Practise safer sex.
- Take up cancer screening opportunities.
- Be safe on the roads: follow the Highway Code.
- Learn the First Aid ABC - airways, breathing, circulation.

The point is not whether or not this is sound advice (others might suggest an alternative ten tips including the injunction not to be poor, unemployed, work in a low paid job, or work in a stressful work-place). The importance lies in the way that it finds new ways of focusing public policy on what was previously a private realm in the name of reducing risk. Citizens are invited to re-schedule their priorities today in the light of a future re-defined by risk and uncertainty. The social objective of the welfare state was to provide citizens with some guarantee of future security. The social objective in a risk society, is to warn of future dangers.

This is also an inclusive agenda. Individuals and groups who are 'hard to reach' should no longer be beyond the reach of the benefits (and responsibilities) of such policies. So paragraph 9.36 in *Our Healthier Nation* asserts that:

> Statutory organisations are working increasingly with individuals, families and communities from black and minority ethnic groups to understand diversity, the different cultural traditions and the various ways in which people from those communities express themselves. For example, health authorities and community organisations are working in mosques, gurdwaras and temples to set up health services including screening services. In this way the local communities have more say in the organisation and delivery of such services.

However, although all of this points us towards a radically transformative discourse - one which changes the nature of the public-private relationship, politicises the self, de-bureaucratises the public sector and so forth - there are limits to the effectiveness, or reach, of the discourse itself. Individuals do not have the Chief Medical Officer's 'ten tips' posted in their local pubs and nor do they send back their pints of beer and demand lightly steamed vegetables instead. There have been no reported cases of marital disputes ending abruptly as the couple recalled the CMO's injunction to 'talk things through'. Equally, voluntary bodies have their own agendas which they will seek to pursue within partnerships. Finally, different parts of the public sector have different constituencies, different statutory requirements, and different cultures. There is every reason to suppose that these will set limits to the pursuit of 'joined-up government'.

Faced with the limits to this discourse, we have seen a variety of attempts to align public and voluntary agencies, to effect cultural change, and to encourage different activities. This includes the use of manufactured uncertainty and futures-oriented thinking. Let us look at some of the techniques being used to achieve this.

Tools to Stabilise the Future

'Tools to stabilise the future' are not intended to allow us to pre-determine those things which will happen in the future. Rather, they describe those intellectual and managerial techniques which transform the infinite possibilities of the future into a manageable landscape in which decision-takers can act. Much futures-oriented thinking in the healthcare sector is explicitly designed to achieve this. At a general level, one important means for stabilising the future is the use of metaphors and we start with this.

When Bevan 'nationalised' British hospitals the dominant metaphor of the policy process was that of the 'machine'. For example, throughout the latter stages of the second world war, the Machinery of Government Committee sat to determine the shape of government after the war. Coinciding with a changed view of the future, a changed view of society and politics emerged. For example, Paul Ormerod (1998) has argued for viewing the political economy as a complex adaptive system. The Nobel Prize-winning American economist, Douglass C. North (1999) similarly argued that we need to understand the complexity of economic change and shake off simplistic accounts of causality. Geoff Mulgan (1997), who was later to join Blair's Policy Unit, argued in 1997 that increasingly dense and complex inter-connections raised new moral and practical problems for policy makers. These, and many others who have influenced the 'new Labour' project, are all united in questioning the 'machine' metaphor of policy implementation.

However, what metaphor should replace the machine? Morgan (1986) argues that shared metaphors are deeply important in organisational life and offers a range of possible metaphors to draw upon: organisations as machines, organisms, brains, cultures, political systems, psychic prisons, flux and transformation, domination. During the 1980s and early 1990s there was a neo-liberal project to escape from the machine metaphor and replace it with a market metaphor. This never successfully established itself and more recently we have, if any pattern can be discerned, moved towards an ecological metaphor; the learning organisation as part of an adaptive and inter-connected system (see Dick, 1999). The parallel development in social theory reflecting this trend has been in the application of social learning theory (see Greener, 1999).

The machine metaphor implies a process conception of time. It is one of inputs, activities and outputs. Time is measured by the policy cycle, feed-back loops, and the gap between implementation and outcome. As the machine metaphor has been used less, and as academics have noted a shift from government to governance, a number of techniques have emerged in the healthcare sector to develop, share and stabilise a new understanding of the temporality of the organisation. This is not necessarily made easier by an ecological metaphor which is cyclical and punctuated by a great variety of timeframes, planning horizons and policy processes. Some of these techniques might be found on a well-known spectrum of futures methods reproduced in Fig 12.1 below.

During the past twenty years, the growing sense of uncertainty has encouraged a shift to the left of the diagram away from methods concerned with achieving certainty and towards methods which embrace ambiguity. However, and importantly, faced with the pressures of the electoral cycle and populist demands over alleged and real medical failures, governments find themselves constantly drawn back into an attempt to establish protocols and rules. Therefore we have the National Institute for Clinical Excellence, Clinical Governance procedures, the Commission for Health Improvement and so on. Consequently, governments not only follow Beck by emphasising that all we can do is gauge uncertainty but also claim to be able to minimise and manage those risks. Consequently, there is a need for techniques which could incorporate profound ambiguities in the decision-taking process. To be clear, this ambiguity is linked both to uncertainty about the wider environment and ambiguity about the nature of the governmental project.

Futures methods

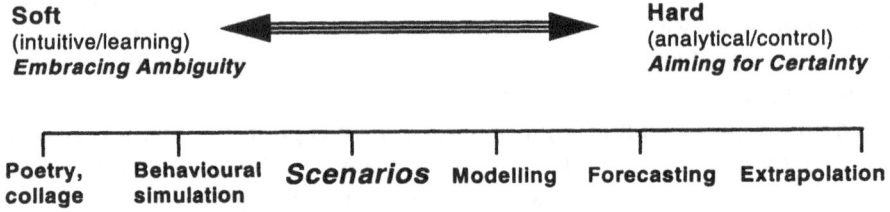

Fig 12.1 Methodological developments in futures thinking

To provide a sense of what these techniques involve, we will briefly consider the use of scenarios and then simulations. Scenario thinking involves using imaginary accounts of feasible futures as an aid to current decision-taking. There are three main forms. The first is a precautionary model. This was used, for example, in the final stages of the Apartheid regime in South Africa to demonstrate the negative future consequences both of continued minority rule and of an unsustainable radicalism. It played a part in aligning the spectrum of opinion in South Africa behind a strategy of universal enfranchisement and restrained redistribution (see Kahane, 1992). A second is the sort often used by environmentalists. We might call this the visionary model. In this a vision of the preferred normative future is outlined (sustainable, consensual, diverse, etc.) and then as time is reversed from the preferable future back to the present day, the steps needed to reach this future can be identified.

In the British healthcare system, however, by far the most common use of scenario thinking is what we might call the learning model. In this approach, between two and four equally desirable and equally plausible futures are described which are based upon a systematic analysis of current trends. This is an approach pioneered in Britain by companies such as Shell and developed further by companies such as ICL (see Davis, 1999;

Ringland, 1997). In the first instance, it involves scenario building through which participants may come to understand the complex inter-connections in their policy arena. Then the process involves using these scenarios either to test existing strategies (the so-called wind-tunnel approach) or to create new policy options (the so-called generator option).[2] These are explained in more detail in Van der Heijden (1997) who emphasises that the purpose is to create a learning, conversational organisation. Because the resistance to learning is emotional as well as intellectual, these processes typically involve the use of actors, colours, video, role play and so forth (see Ling, 1999b). Above all, they seek to use strong metaphors to allow complex visions of the future to be quickly shared.

However, each of these scenario-based techniques involves the participants in a complex reflexive process. The process simultaneously challenges prevailing assumptions and requires participants to more deeply reflect on their understanding of the causal processes and the values which might underpin their actions. In the last two years there have been over a dozen scenario sets for the British health system, and these have been associated with hundreds of workshops. This suggests, at the very least, a great deal of reflexive thinking (although measuring the impact of this is very difficult).

Alongside scenario techniques, there is the use of simulations. The term is applied to a variety of rather different activities. These include the application of systems-modelling and the interactive use of computer models (such as Systems Dynamics). Here we are discussing what Philip Hadridge describes as 'group simulation' (see Hadridge, forthcoming). This is less about exploring the wider environment (used in scenarios) and more about exploring the immediate system within which the participants work. A simulation will typically set up the rules of the game (say, the delivery of mental health services in a particular locality) and then have participants take on the various roles within that system. Simulated time might then move forward at a rate of, say, six months every half-hour and the participants will have various plausible problems thrown at them. The purpose is to test the system and to rehearse reactions to possible events in order to improve decision-taking. It is also to engage the imagination so that participants 'live' the future. For this reason, simulations will often have badges, team T-shirts, pretend press conferences and other means to engaging the imagination. Once again, the process involves an iterative reflection with various colleagues about the ways in which decisions might be made.

Scenarios and simulation are both responses to the growing sense of uncertainty and the way in which 'risk' changes in the behaviour of individuals and organisations. Uncertainty and risk bring the future into organisations in new ways and these have produced particular responses in the healthcare sector. These responses recognise the emotional as well as the intellectual consequences of risk and uncertainty. To address the emotional consequences of uncertainty, futures-oriented thinking offers new metaphors and narratives which can be shared within organisations. These replace the older narratives of progress and the policy machine. They also seek to address the intellectual needs through the use of other techniques. These include the horizon scanning exercises of the Office of Technology's Foresight Programme and non-governmental initiatives such as the Nuffield Trust/Judge Institute's 'Policy Futures for Health' project. They also introduce highly complex notions of systems modelling and policy ecosystems. Together, this is an attempt to create a new type of organisation which can respond in a more reflexive and learning way which reflects the changing 'eco-system' of health organisms. If this is so, who is owning and driving this process?

Owning the Future: Some Protocols for Understanding Health Futures

In Bosch's Millennium Triptych Berger saw a particular image of the future in which space and time were disordered, boundary-less and stripped bare of meaning. We have seen that this captures an aspect of late modernity. We can see this in increasing complexity and interconnectedness, decreasing predictability, the impulse towards de-bureaucratisation and the rejection of the machine-metaphor as the basis for organisational life. However, we can also see that in this context organisations in healthcare have sought to stabilise the future through developing new techniques. It is self-evidently not the case that post-modernisation has created complete and radical uncertainty. By highlighting the parameters of uncertainty and risk, knowledge points to a shifting and contested future and not to the end of history. For this reason Castells is wrong to suggest that 'timelessness is the recurrent theme of our age's cultural expressions...' (Castells, 1996). In the organisational cultures of the healthcare sector, at least, a quite different theme can be found. This is the theme of how to construct and then stabilise an account of the future, how to align organisations behind such a project, and how to generate actions and organisational change based upon this shared vision. Understanding this process is fundamental to understanding the changing disposition of power and the emerging capacities of the public sector. We do not have the space to answer all the questions raised by this theme but we can at least establish some protocols for understanding it and indicate some highly provisional conclusions. In doing so, I will draw heavily upon my own informal knowledge of this sector based upon extensive consultancies in health futures during the past three years. Clearly, this would need to be analysed in a more systematic fashion before drawing firm conclusions. What is presented here, therefore, is an emergent project for future research.

Who Pays For and Commissions Futures Work?

One of the largest futures oriented exercises in recent years in Britain is the Government's Foresight programme. Since its formation in 1994 it has sought to align businesses and public organisations (including Departments and Research Funding Councils) behind a shared vision of possible futures and agreements about future needs. The Department of Health has also taken an active interest in the possibilities of scenario work, sponsoring in 1998, for example, a workshop involving both senior civil servants and leading figures from the pharmaceutical sector. The NHS Confederation, also in 1998, supported one of the largest scenario-building exercises in the British healthcare sector (the so-called Madingley Scenarios), sharing the costs with the Nuffield Trust. In 1999, the Nuffield Trust also supported both the so called 'Pathfinder' Report on Health Futures (Dargie, 1999) and the Public Health Genetics Unit work on the impact of developments in human genetics on future public health action and clinical services. A sense of the range of work being produced by practitioners (rather than academics) relevant to the future of British healthcare can be gauged from Fig 12.2 (see below).

British Medical Association (1995) *Future models for the NHS: a discussion paper*, British Medical Association, London.

Coile, R. (1995) Health buildings - the next generation, *Hospital development*, 26, 2, pp.12-14.

Coote, A. and Hunter, D. (1996) *New agenda for health*, Institute for Public Policy Research, London.

de Geus, A. (1997) *The Living Company*, Harvard Business Press.

Ferguson, T. (1997) Healthcare in cyberspace: patients lead a revolution, *The Futurist*, Nov/Dec.

Hadridge, P. *et al.*, (1995) Tomorrow's World, *Health Service Journal* 5/1.

Harrison, A. and Prentice, S. (1996) *Acute futures*, Kings Fund, London.

House of Commons Select Committee on Health, (1996) *Long-term care: future provision and funding*, HMSO, London.

Ling, T. (1998) 'Future Issues' in NHS Confederation/Policy Information network, *Your Finger on the Pulse*, NHS Confederation.

Office of Public Management, (1995) *Public services 2007*, Office of Public Management, London.

Sapirie, S. (1994) What does 'health futures' mean to WHO and the world? (Introduction), in *World Health Statistics Quarterly*, World Health Organisation, Geneva.

Sapirie, S. and Orzeszyna, S. (1995) *Health Futures. The Results and Follow-up of the 1993 Consultation*, World Health Organisation, Geneva.

Schreuder, R. (1995) Health scenarios and policy-making: lessons from the Netherlands, *Futures*, 27, p.9.

Schwartz, P. (1991) *The art of the long view*, Doubleday and Currency, New York.

Senge, P., Kleiner, A., Roberts, C., Ross, R.B. and Smith, B.J. (1994) *The Fifth Discipline Fieldbook*, Doubleday.

Shoemaker, P. and van der Heijden, K. (1992) Integrating scenarios into strategic planning at Royal Dutch / Shell, *Planning Review*, 20, 3, pp.41-46.

Smith, R. (1997) Editorial, The Future of Healthcare Systems, *BMJ*, 314.

Van Der Heijden, K. (1997) *Scenarios: the Art of Strategic Conversation*, John Wiley, Glasgow.

World Health Organisation, (1994) *Health futures in support of health for all* - Report of and International Consultation convened by the World health Organisation, Geneva, 19-23 July 1993. WHO/HST/93.4.

Fig 12.2 Some examples of futures work in use in British healthcare

One thing is fairly clear about the funding and commissioning of futures work around healthcare; financial and political support for it is coming from very senior levels within the health community. This is paralleled by similar (although less public) scenario work in companies close to the

NHS. These people appear to have bought into the idea that uncertainty is growing and the pace of change quickening. They clearly perceive that they might need to do things differently. However, a discourse analysis of the texts in Fig 12.2 would reveal, I suspect, a fairly clear sense of how we are moving into uncharted waters, how this requires new cultures and organisations, and how this in turn requires a new approach to leadership. However, as we can see below, this discourse has been used in selective ways.

Who Carries Out Futures Work?

In recent years there has been increased academic interest in the field of futures in general and health futures in particular. Often this has been associated with business schools (the Judge Institute at Cambridge University, Nottingham Business School, Leeds Metroplitan) and sometimes with a broader strategic view in mind (the Science and Technology Studies Unit at York University and David Mercer's 'Futures Observatory' from the Open University). More directly involved with practitioners, however, are consultancies such as the Office of Public Management, Idon Associates and State of Flux. Others act almost as 'internal consultants' from within the NHS and Department of Health such as Richard Walsh, Geoff Royston, Peter Dick, and Philip Hadridge.

Those most directly involved in carrying out futures work from within healthcare have been influenced by a particular part of the literature which can be seen in Fig 12.2. In particular, at least most of these have been influenced by the work by Van der Heijden, de Gues, Senge, and Schwartz. This is a body of work that stresses the importance of organisational learning, and of workshops intended to stimulate good 'conversations'.

Who is Involved?

In building scenarios, participants have been people with a view of the whole system and a claim to expertise in at least part of it. Almost inevitably, this has meant relatively senior people from within the area being considered. In an attempt to balance this, scenario building projects also seek out 'exceptional' views which might add a different dimension to the views of the experts. In simulations, the requirements are rather different in that they typically involve the actual decision takers from within the system being simulated, often acting out their 'real' roles.

How is the Work Disseminated?

My own view is that there is a strange paradox at work here. All of the futures-oriented work I have seen in and around the healthcare sector in recent years has emphasised the need for flexible, bottom-up responsive, partnership-making behaviour which is owned by people working at every level within the organisation. Leaving to one side whether or not this is indeed the correct conclusion, it is at least strange that to date, futures work has very rarely drifted 'down' the system to the grass-roots level. Rather, futures work has been regarded as the property of established key decision-takers. By implication, at least, its practical purpose appears to be to create a more informed elite so that they can pursue better policies rather than to achieve the organisational learning objectives espoused by the intellectual gurus of future work. In other words, the aim may be seen to be to strengthen the existing structures rather than to transform these structures in the face of an uncertain future. This may not be tenable.

Conclusions

There are compelling reasons for believing that decision-takers now believe that that the future is less predictable, that the pace of change is quickening, and that knowledge illuminates risks rather than certainty. It is even possible that this belief reflects real changes taking place in late modernity. Consequently, the intellectuals of this new era announced that we need to embrace complexity, forego bureaucracy, and use inter-connectivity to achieve our goals. These ideas have fed into the so-called Third Way thinking. However, my provisional conclusions suggest that when these ideas are reconstituted and fed into the power structures of existing organisations they are used selectively to defend the existing power structures. These are the power structures of a state bureaucracy. This should not surprise us. There is always a gap between the new technology (in this case the socio-technology of scenario planning) and its application. However, this should be seen as opening up a question for further analysis and as an invitation for future research and the conclusions arrived at should be treated as tentative. However, given the absence of any systematic study of this question, it is at least a start.

There is in UK health policy, then, a complex set of relations connecting the perception of increasing uncertainty, the quantification of risk, and the emergence of futures-oriented thinking. However, the particular way in which futures thinking has been commissioned, conducted and used has also been shaped by the wider disposition of power within the health sector.

The result is that futures oriented thinking has been more about ensuring that bureaucratic and political leaderships are not 'caught out' by an unanticipated turn of events and less about building a more responsive, flexible and competent healthcare system.

Notes

1. Private communication with Richard Smith, Editor of the British Medical Journal, December 1999.
2. The Government's Foresight programme was established in 1994. Its self-professed aim is to enhance the quality of life by:

- Establishing visions of the future and identifying priorities for action to help the UK meet its future needs;
- Developing a culture of forward-thinking about market technology opportunities and threats;
- Creating enduring networks linking business, the science base, and government as a basis for generating action on the priorities identified.

Prior to publishing its first reports in 1995 the programme consulted with 10,000 people and the reports themselves aimed to identify the likely social economic and market trends in each sector over the next 10-20 years. They also intended to identify necessary developments in science engineering, technology and infrastructure if future needs were to be met. The programme also seeks to feed its insights into programmes and awards designed to encourage more effective (pre-competitive) collaboration; improved foresight practices in firms; the programmes of other Departments, Research Councils, Learned Societies and so on. (See Office of Science and Technology, 1998.)

References

Appleyard, B. (1999) *Brave New Worlds. Staying human in the genetic future*, Harper Collins, London.

Beck, U. (1992) From industrial society to the risk society - questions of survival, social structure and ecological enlightenment, *Theory Culture and Society*, 9.

Berger, J. (1999) Welcome to the abyss, *The Guardian*, November 20th.

BMJ (1997) The future of healthcare systems, *British Medical Journal*,314, p.70009.

Castells, M. (1997) Hauling in the future, *The Guardian*, December 13th.

Coote, A. and Hunter, D. (1996) *New agenda for health*, Institute for Public Policy Research, London.

Dargie, C. (1999) *Policy Futures for UK Health. A Consultation Document*, The Nuffield Trust, London.

Davis, G. (1999) Foreseeing a refracted future, *Scenario Strategy and Planning*, April/May, 1,1.

Department of Health (1998) *A First Class Service. Quality in the new NHS*, Department of Health, London.

Dick, P. (1999) *Public Ecosystems: metaphor and framework*, unpublished report arising from his Fulbright-Humphrey Fellowship (1998-99).

Du Gay, P. (1996) Organizing Identity, in Hall, S. (ed.) *Questions of Cultural Identity*, Sage, London.

Greener, I. (1999) *The Politics of the Policy-making Process in the NHS: A social learning perspective*, Unpublished PhD thesis, Anglia Polytechnic University.

Hadridge, P. (forthcoming) *Learning through group simulation*, www.cambridge forsight.com.

Harrison, S., Hunter, D. and Pollitt, C. (1990) *Just Managing: Power and Culture in the NHS*, Macmillan, London.

IBM, (1997) *The Net Result*, Report of the INSINC Working Party.

Kahane, A. (1992) The Mont Fleur scenarios: *Weekly Mail*, Johannesburg, and *Guardian Weekly*, London and Manchester.

Kevles and Hood (1992) *The Code of Codes*, Harvard University Press, Cambridge, Mass.

Klein, R. (1995) *The New Politics of the NHS* (Third Edition), Longman, London.

Ling, T. (1999a) Which way to a health future? Reflections on the Madingley scenarios, *Foresight*, 1,1, February.

Ling T. (1999b) Community Foresight, *Scenario and Strategy Planning*, 1, 3.

Ling, T. (1994) The new managerialism and social security, in Clarke, J., Cochrane, A. and McLaughlin, E., *Managing Social Policy*, Sage, London.

Ling, T. (forthcoming) Unpacking partnerships in healthcare, in Clarke, J. et al., *New Labour New Welfare?* Sage, London.

Marinker, M. (1998) Looking and leaping in Marinker, M. and Peckham, M., *Clinical Futures*, BMJ Books, London.

Morgan, G. (1986) *Images of Organization*, Sage, London.

Mulgan, G. (1997) *Connexity: how to live in a connected world*, Harvard Business School Press, Cambridge, Mass.

North, D.C. (1999) *Understanding the Process of Economic Change*, Institute for Economic Affairs, London.

Office of Science and Technology (1998) *Foresight. Consultation on the next round of the Foresight programme*, Department of Trade and Industry, London.

Ormerod, P. (1998) *Butterfly Economic*, Faber and Faber, London.

Osborne, D. and Gaebler, T. (1992) *Re-inventing Government*, Addison Wesley, Reading, Mass.

Ringland, G. (1997) *Managing for the Future*, John Wiley, London.

Rothman, H.(1998) Foreword in Wheale, P. Schomberg, R. von and Glasner, P. (1998) *The Social Management of Genetic Engineering*, Ashgate, Aldershot.

Van Der Heijden, K. (1997) *Scenarios: the art of strategic conversation*, John Wiley and Sons, Glasgow.

Wilkie, T. (1993) *Perilous knowledge. The human genome project and its implications*, Faber and Faber, London.

Index